Ever Rojas Huamán
Christian Edinson Murga Tirado
Rolando Fuentes Llave

One-way capacity and service levels, based on HCM methodology

Ever Rojas Huamán
Christian Edinson Murga Tirado
Rolando Fuentes Llave

One-way capacity and service levels, based on HCM methodology

An analysis from the point of view of the HCM methodology

ScienciaScripts

Imprint

Cover image: www.ingimage.com

This book is a translation from the original published under ISBN 978-620-3-87832-5.

Publisher:
Sciencia Scripts
is a trademark of
Dodo Books Indian Ocean Ltd. and OmniScriptum S.R.L publishing group

120 High Road, East Finchley, London, N2 9ED, United Kingdom
Str. Armeneasca 28/1, office 1, Chisinau MD-2012, Republic of Moldova, Europe
Managing Directors: Ieva Konstantinova, Victoria Ursu
info@omniscriptum.com

Printed at: see last page
ISBN: 978-620-8-84389-2

Contents

DEDICATION

I dedicate this Thesis to our Almighty God the Father, who opens roads no matter how difficult and rough they may seem; to my beloved daughters Anggie and Nataly, who are my strength in the face of the goals I have set for my life; to my beloved wife Eliana for her positive influence on my decisions and to my parents, whose smiles show their happiness for my achievements.

THANK YOU

My most sincere thanks to the Universidad Privada del Norte, for giving me the opportunity to carry out this second career; to my advisor Alejandro Cubas Becerra, for his guidance in the development of the Thesis; to Jorge Ancajima Carrasco and Christian Murga Tirado, for their decisive support in the collection of data.

SUMMARY

The main objective of this research thesis is to analyse the capacity and levels of service of Mario Urteaga Avenue, section: Jr. Dos de Mayo to Óvalo el Inca, using the recommendations of the Highway Capacity Manual: HCM 2010, of the Transportation Research Board of the National Academies of the United States (TRB). The results obtained for one week of study, with vehicle capacity calculations of 14.5 hours per day were: For the road segment, a capacity of 1532 veh/h and level of service "E"; for the traffic-light intersection (Av. Mario Urteaga - Jr. Urrelo), the capacities of: 1136 veh/h (SE access), 328 veh/h (NW access), 452 veh/h (SW access) and 552 veh/h (NE access), respectively with levels of service "F" both separately and jointly.

Keywords: Capacity, level of service, HCM 2010.

CHAPTER I

INTRODUCTION

1.1. Problem reality

The problems observed in traffic engineering, which have to do with road traffic, are diverse, and need to be addressed from a technical-scientific perspective using multimodal methodologies and procedures, which allow an objective evaluation of traffic conditions and road service quality indicators based on the user's perception. One of the critical problems that road planners have to deal with is traffic congestion, a problem that has been increasing in most parts of the world, with a tendency that will continue as cities grow towards becoming large metropolises, and as the economic conditions of nations achieve stability. Presumably, the problem of traffic congestion leads to other types of problems such as: delays, increased pollution, higher fuel consumption, increased stress levels, etc.

Quintero G. (2017), in his scientific article: "From the concept of traffic engineering to sustainable urban mobility", mentions the complexity of the problem and the solution approaches it requires. He states that the treatment of mobility in cities based on specialised studies of variables such as traffic volume, speed and density, in addition to other aspects such as infrastructure inventories, generation of forecast models, modelling with the use of software, etc., although it is true that they have facilitated the development of a new approach to mobility in cities, Although it is true that these studies have greatly facilitated the characterisation of traffic behaviour, as well as the design of measures for its improvement, oriented towards the planning of traffic operation and the physical intervention of transport infrastructure, it is nevertheless mentioned that in practice, these studies have left aside the analysis of essential factors such as social equity, environment and economic value, typical of the conception of the problem of urban mobility within the framework of the sustainability of transport systems.

Worldwide, we can list colossal cases of traffic congestion, such as the one in China in August 2010 (considered the biggest traffic jam in history). On the G110 motorway, which links Tibet and Beijing, there were 110 kilometres of congestion due to two factors that, if combined, could paralyse half the country: roadworks and operation return. The congestion was of such magnitude that it lasted fifteen days and some of the vehicles were moving at a rate of one kilometre per day (Ortiz, 2016).

In the Latin American case, Fernández De La Maza (2018), mentions that two of the 10 most congested cities in the world are in Latin America, according to a traffic score analysis; referring to Sao Paulo and Bogotá.

In its version of 3 September, the newspaper el Comercio, Macera (2018), mentions that the government would save US$500 thousand if it manages to improve the existing traffic chaos in the city of Lima. Approximately 7,000 vehicles travel daily on the Carretera Central del Perú, a figure that far exceeds the maximum capacity for the route, 4,000, according to figures from the Ministry of Transport and Communications (MTC). The panorama, however, becomes more serious when future estimates are analysed: by 2035, the road will support the weight of 14500 vehicles per day, i.e. it will demand 362% of its capacity. The problem is exacerbated by the lack of comprehensive strategies that can be implemented by the current administration.

In the local case, Cajamarca is not exempt from the problem of vehicular traffic, mainly due to the oversupply of public transport vehicles and cars, taxis and motorbike taxis, lack of parking control, lack of vertical and horizontal signalling, lack of infrastructure for public transport such as bus stops, etc., which are accompanied by a lack of authority for the planning, supervision and administration of the organisation and operation of the public transport system.

This problem, and others, typical of traffic engineering, requires an exhaustive study in each case; one of the starting points is the analysis of the capacity and service levels of the urban and rural transport routes .

Regarding the background considered in this research, we begin by referring to Sabando Santibáñez,

(2017) and his study on determining the level of service on urban streets. Based on the Highway Capacity Manual 2010, he developed a practical guide adapted to the national reality, establishing a series of criteria necessary to determine the level of service in urban streets. These criteria were based on performance measures that are measurable on the ground and perceptible by travellers, such as travel speed and delay, which reflect the factors that influence travel time along a street segment and the delay that occurs at signalised intersections. For the Fermín Vivaceta urban road axis in the city of Santiago, it resulted in a level of service that reflects a stable performance of the analysed road facility. However, when considering an individual analysis by section, it was detected that a segment of the analysis was operating under unacceptable service conditions.
unacceptable; therefore, it must be considered in the overall context of the interpretation of the level of service.
Cuenca Poma (2014) carried out an analysis of the capacity and levels of service on the Loja Zamora road, based on the following parameters: traffic volume, speed and geometric characteristics of the road. The road under study, characterised by two lanes, connects the cities of Loja and Zamora in southern Ecuador over a length of 60.7 km. The geometric variables established for the road were: length, average gradient, length of passing zones and density of access points. Traffic and speed information was determined using a pneumatic type vehicle counter for a 24-hour record.
At the national level, Agreda Ríos & Parra Alarcón (2017), present a study focused on a comparative analysis of the application of the HCM 2000 and HCM 2010 methodology in a traffic-light section of San Luis Avenue in the metropolitan city of Lima. The general objective of this study was to compare the results obtained from the application of the HCM 2000 and HCM 2010 methodologies in a series of traffic-light intersections and to define which of them is more applicable to the reality of traffic in Metropolitan Lima. It managed to find some incompatibilities in terms of the average blocking time of buses per stop: according to the HCM in both versions, the blocking time per stop of buses on green is 14.4 seconds. However, according to the analysis made from own experience, the time per bus stop at one of the intersections does not exceed 9 seconds. Thus the bus blocking adjustment factor (f_{bb}) increases by 2.4%. While this variation is not very considerable, it can make a difference in finding the level of service and increases the value of the adjusted saturation flow rate. HCM then proposes a conservative formula to find the blocking factor per bus.
On the other hand, Chuquihuaccha Mauricio & Galván Aldoradín (2016), provide a proposal to improve vehicle congestion using HCM 2010 methodology at the intersections of 26 de noviembre Avenue, between Pachacútec Avenue and Salvador Allende Avenue, in the district of Villa María del Triunfo. Through the traffic signal redesign at one intersection it was able to improve from a level of service D to a level of service B, implying that the existing traffic light timings are not optimal and that the improvement of the intersection would not imply an expense in geometric modifications. For the intersections analysed, the theoretical calculation of the levels of service is validated and reflects results similar to those obtained in the simulation programmes. Therefore, they could continue to be used to consider other improvement proposals not considered in the work carried out.
In turn, Arguedas Bemal & Mosqueira Grosso (2018), present a proposal for a comprehensive solution on Del Aire Avenue between Aviación, San Luis and Rosa Toro Avenues, in the city of Lima, through an analysis of vehicular congestion by applying the HCM 2010 methodology. As a result of their research, they were able to improve the level of service at the intersection of Aviación and Del Aire Avenues from a level "E" to a level "D" by: a) modifying the existing traffic light cycle and b) redirecting heavy vehicles on one of the avenues.
At the local level, Alcántara Quispe (2018) carried out an analysis of the level of service and vehicular capacity of San Martín de Porres Avenue located between Atahualpa Avenue and Argentina Avenue, applying the methodology of the HCM 2010. To carry out the analysis of the level of service, the road was divided into segments taking into account the direction of vehicular traffic: segment I (vehicular traffic from Av. Atahualpa to Av. Argentina) and segment II (vehicular traffic from Av. Argentina to Av. Atahualpa). A daily gauging was carried out from 7:00 a.m. to 8:00 p.m. with 5-minute intervals

during one week; it was determined that for both segments the day with the highest volume of vehicles is Monday and the hour with the highest volume of vehicles is between 6:30 p.m. and 7:30 p.m. (segment I), and between 6:00 p.m. and 7:00 p.m. (segment II); likewise, the 5-minute interval with the highest volume of vehicles identified was between 6:45 p.m. and 7:00 p.m. The volume for the 15 minutes of peak demand, from 6:45 pm to 7:00 pm, was 400 vehicles for Segment I and 339 vehicles for Segment II. The peak hour volume was 1231 vehicles for segment I (6:30 pm to 7:30 pm) and 1240 vehicles for segment II (6:00 pm to 7:00 pm). Subsequently, by applying the equations established by the method, it calculated the base free flow speed and the displacement speed, obtaining that the base free flow speed is 41.15 mi/h (66.22 km/h) for segment I and 41.20 mi/h (66.31 km/h) for segment II. Likewise, the displacement velocity obtained was 14.68 mi/h (23.62 km/h) for segment I and 11.69 mi/h (18.81 km/h) for segment II. He then calculated the percentage ratio between the base flow velocity and the travel velocity to determine the level of service. One of the conclusions obtained in this work was that the section of Av. San Martín, located between Av. Atahualpa and Av. Argentina, has a variable level of service due to the fact that segment I has a level of service E and segment II has a level of service F, with a vehicle capacity of 1600 vehicles/h and 1356 vehicles/h respectively in each segment.

Finally we mention Llanos Rimarachín (2018), who carried out the analysis of the level of service of the traffic-light intersections with the highest traffic flow on Hoyos Rubio Avenue. For the development of the methodology, data was collected using the traffic video-recording tool, which was applied at each access to the intersections under study for a period of one week. Topographical surveys were also carried out to establish the geometric, traffic light and traffic conditions, which were essential for the application of the HCM methodology to determine the level of service based on the average delay per vehicle using spreadsheets. The delay for the first intersection was 353.7 seconds and for the second, 191.7 seconds, which means that both intersections have a level of service F. This level is considered unacceptable and shows problems in vehicular traffic, so it is imperative to consider possible alternative solutions. The proposed improvements decreased these delays to 43.5 seconds and 40.4 seconds respectively at each intersection, optimising their serviceability to a level of service D.

The conceptual definitions considered in this research to support the research problem have been divided into conceptual definitions for the analysis of capacity and levels of service at roadway segments; and conceptual definitions for the analysis of capacity and levels of service at signalised intersections. These are set out below:

TRANSIT ENGINEERING, referring to that phase of transport engineering which deals with the planning, geometric design and operation of traffic on streets and highways, their networks, terminals, adjacent lands and their relationship with other modes of transport (Cal, Reyes Espíndola, & Cárdenas Grisales, 1995).

The BASIC ELEMENTS OF TRANSIT ENGINEERING are: the user (pedestrians and drivers), the vehicle and the road (urban roads and highways) (Cal, Reyes Espíndola, & Cárdenas Grisales, 1995).

The USER is made up of PEETONS, who are those people who travel on foot on public roads (they are also considered to be those who push any other non-motorised vehicle of small dimensions, and those with reduced mobility who circulate on foot with a wheelchair); and DRIVERS, who are the people in charge of driving a motor vehicle for the transport of people, animals and others.

The VEHICLE is the means of locomotion that makes it possible to move people or things from one place to another.

The VEHICLE CLASSIFICATION; it is regulated by the National Regulation of Vehicles of the Ministry of Transports and Communications MTC (2003, P. 54,55); which according to the SUPREME DECREE N° 058-2003-MTC, establishes the requirements and technical characteristics that the vehicles must fulfill so that they enter, transit, operate and retire of the

National System of Terrestrial Transport. It takes for north the protection and security of the people, environment and safeguard of the road infrastructure:

- **Category L.-** Motor vehicles with less than four wheels.

Ll: Vehicles with 2 wheels, up to 50 $cm^3$and maximum speed of 50 km/h.
L2: 3-wheeled vehicles, up to 50 $cm^3$and a maximum speed of 50 km/h.
L3: 2-wheeled vehicles, over 50 cm^3or speed over 50 km/h.
L4: Vehicles with 3 wheels asymmetrical to the longitudinal axis of the vehicle, more than 50 cm^3or a speed greater than 50 km/h.
L5: Vehicles with 3 wheels symmetrical to the longitudinal axis of the vehicle, over 50 cm^3or speed over 50 km/h and whose gross vehicle weight does not exceed one tonne.

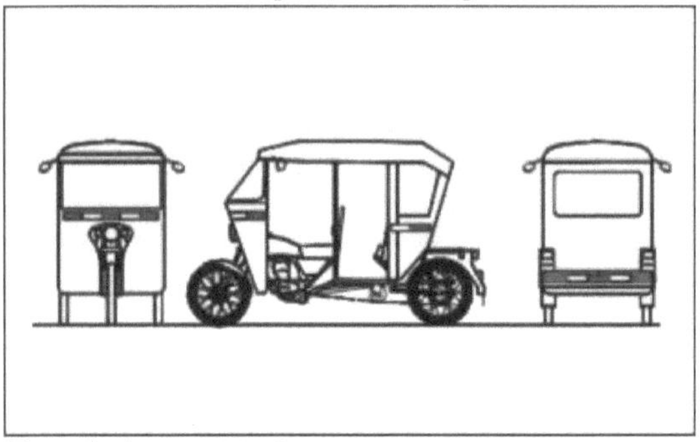

Figure 1 Trimoto, vehicle of category L5.
Source: MTC National Vehicle Regulations 2003, P. 71.

- **Category M.-** Motor vehicles of four wheels or more designed and constructed for the carriage of passengers.

for the carriage of passengers.
MI: Vehicles with 8 seats or less, excluding the driver's seat.

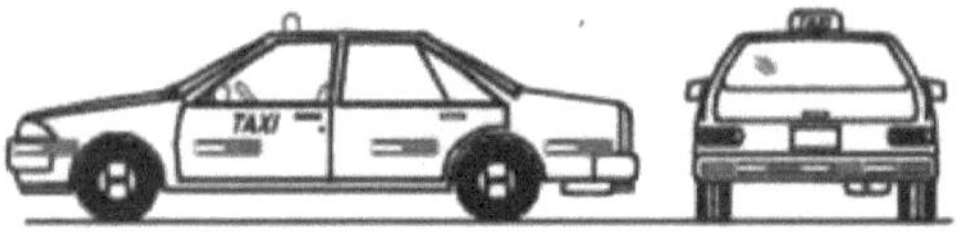

Figure 2 Automobile or Taxi, vehicle of category MI.
Source: MTC National Vehicle Regulations 2003, P. 71.
M2: Vehicles with more than 8 seats, excluding the driver's seat and gross vehicle weight of 5 tonnes or less.
M3: Vehicles with more than 8 seats excluding the driver's seat and a gross vehicle weight of more than 5 tonnes.

Figure 3 Buses, minibuses and minibuses, vehicles of categories M2 and M3.
Source: Taken from MTC National Vehicle Regulations 2003 2003, P. 72.

According to the arrangement of passengers, vehicles of categories M2 y M3, are classified into:

Class I: Vehicles constructed with standing passenger areas allowing frequent passenger movement.

Class II: Vehicles constructed primarily for the carriage of seated passengers and also designed to permit the carriage of standing passengers in the gangway and/or in an area not exceeding the space provided for two double seats.

Class III: Vehicles constructed exclusively for the carriage of seated passengers.

- **Category N.-** Motor vehicles of four wheels or more designed and constructed for the carriage of goods.

The VIA is that strip of land prepared for the transit of vehicles.

Elements of the URBAN ROAD:

- SIDEWALK: Part of the urban road, intended for the exclusive use of pedestrians or passers-by, and with an appropriately designed elevation with access for the physically handicapped appropriately located.
- ROADWAY or ROLLING SURFACE: A strip of road specially designed for vehicular traffic.
- LANE: Part of the carriageway or road surface, of sufficient width for the circulation of a single row of vehicles.
- DITCHES: Run parallel to the shoulders, intended to facilitate the drainage of the longitudinal surface of the road.
- TRANSVERSAL DRAINAGE: This is formed by culverts and larger structures such as bridges, which allow water to cross from one side of the road to the other without invading its surface.

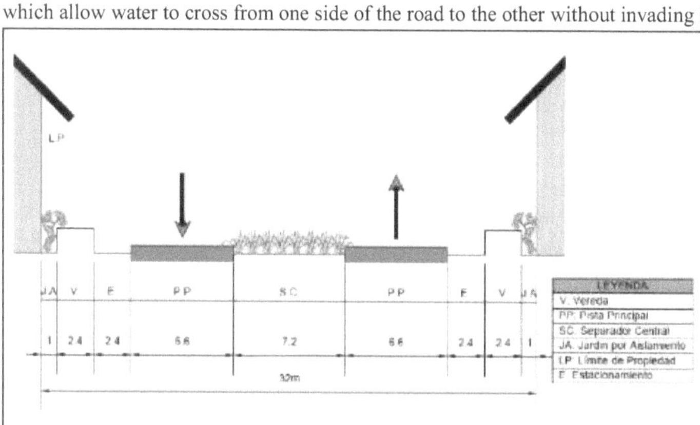

Figure 4 Example of a typical section of a collector road with two lanes for urban developments for residential use.

Source: ICG, Manual de Diseño Geométrico de Vías Urbanas, (2005).
CLASSIFICATION OF URBAN ROADS. The classification follows the principles of specialisation, hierarchy and functionality, and considers: Express, Arterial, Collector, Local and Special Design Roads.
EXPRESS ROADS are used for long journeys. They connect areas of high traffic generation, transporting large volumes of vehicles at high speed and low accessibility conditions, thus facilitating optimal mobility for direct traffic (ICG, Ingeniería de Tránsito y Seguridad Vial, 2009).
ARTERIAL ROADS allow links with medium or high fluidity, low accessibility and relative integration with neighbouring areas. These roads should be integrated to the Express Roads system and allow a good distribution and distribution of traffic to the Collector and Local Roads (ICG, Ingeniería de Tránsito y Seguridad Vial, 2009).
COLLECTOR ROADS facilitate traffic flow from local roads to arterial roads and in special cases connect local roads to express roads. This type of road has often been given the generic name of Jirón, Vía Parque, and even Avenida.
LOCAL ROADS are those whose main function is to provide access to properties or lots, and must carry only their own traffic, both incoming and outgoing. They connect with each other and with collector roads. They are commonly called streets and passages (ICG, Ingeniería de Tránsito y Seguridad Vial, 2009).
SPECIAL ROADS are all those whose characteristics do not conform to the classification established previously. Among them, we can mention the following roads and passages
Pedestrian walkways, boardwalks, promenades, roads that are part of parks, squares or small squares, and roads in tunnels (ICG, Ingeniería de Tránsito y Seguridad Vial, 2009).
Road signs are the physical means used to indicate to road users the correct and safest way to travel on public roads.
The signage, within a road, fulfils the following fundamental functions: organises traffic, warns of dangers, orders safety behaviour, communicates useful information. (Alcántara Quispe, 2018).
Regarding the BASIC PARAMETERS OF VEHICULAR FLOW, in accordance with the HCM 2010, we have:
VOLUME: Total number of vehicles passing over a given section of road during a time interval. This time interval can be variable, but is commonly expressed in terms of annual, daily, hourly, or less than hourly periods.
VEHICLE FLOW: Corresponds to the equivalent hourly rate at which vehicles pass over a given section of road during a time interval of less than one hour, usually 15 minutes.
With regard to SPEED, it should be borne in mind that in order to characterise SPEED, a representative value should be used, due to the wide diversity of individual speeds present in a traffic flow.
The basic speed parameters applicable to a traffic flow on urban streets are as follows:
- AVERAGE TRAVEL SPEED - This is the length of the segment divided by the average travel time of vehicles traversing the segment, including delay times due to stops. (HCM, 2010).
- FREE FLOW SPEED - Defined as the average speed of vehicles on a given segment, measured under low volume conditions, where drivers are free to travel at their desired speed and are not restricted due to the presence of other vehicles or traffic control devices at the downstream intersection (e.g. traffic lights, roundabouts or stop signs) (HCM, 2010). (HCM, 2010).
DELAY is also an important performance measure for elements of a disrupted flow system.
There are several types of delay, where CONTROL DELAY is the main service measure in the HCM for assessing the level of service at signalised intersections. This delay is the delay caused by the presence of a traffic control device at the intersection and includes the delay resulting from the slowing of vehicles arriving at the intersection, the time spent stopping, the time spent moving forward in the queue and the time required to accelerate to the desired speed. (HCM, 2010).
Other types of delay used and considered by HCM are as follows:

- GEOMETRIC DELAY: Delay caused by the geometric characteristics of the track.
- INCIDENT DELAY: Additional travel time experienced as a result of an incident.
- TRAFFIC DELAY: Delay caused by interaction between vehicles, forcing drivers to reduce their speed below free-flow speed.
- TOTAL DELAY: Sum of all the above delays.

CONGESTION INDEX is defined as the ratio between the time spent travelling through the segment and the time spent in free flow, (Congestion index report, 2011), cited in (Alcántara Quispe, 2018):

$$CI = \frac{T_r}{T_l} \qquad (1)$$

Where:

- CI: congestion index.
- *Tv.* Travel time through the segment (sec).
- *TI:* Travel time under free-flow conditions (sec).

The travel time index can be understood by relating the value to a percentage. If the index is 1, then the average travel time is the same as the free-flow travel time, meaning that there is no congestion. If the congestion index is 1.5 or greater it means that the actual travel time is 150% of the free-flow time, or it takes 1.5 times longer to travel a segment than it would under uncongested conditions, therefore, congestion exists. (Alcántara Quispe, 2018)

Taking into account that the relationship between intensity, speed and density is one of the most important aspects in traffic engineering; for the case of this study we refer only to the automobile mode.

CAPACITY represents the maximum sustainable intensity at which people and vehicles can reasonably be expected to traverse a point or cross-section of a lane or road during a given time period and under known traffic, environmental and regulatory conditions. (HCM, 2010).

INTENSITY is a parameter that quantifies the number of vehicles travelling on a point or cross-section of a lane or road during a given time period. Although any time interval can be used, it is typically expressed in hourly terms (i.e. veh/h or veh lig/h), in order to be able to compare and relate quantities (HCM, 2010).

Intensity is a variable that helps to quantify DEMAND, i.e. the number of occupants or drivers (usually expressed as the number of vehicles) wishing to use a given road element during a specific time period (typically 1 hour OR 15 minutes). (HCM, 2010)

The Peak Hour Factor (PHF) is the parameter that estimates the intensity corresponding to the worst period (most loaded period) of the 5 min within it.

In the HCM we find the following relationship to define the FHP:

$$FHP = \frac{intensidad\ en\ el\ conjunto\ de\ la\ hora}{intensidad\ en\ el\ período\ más\ cargado}$$

intensity in the whole hour

intensity in the most loaded period

For periods of 5 minutes, the FHP will be calculated by the equation:

$$FHP = \frac{I_h}{4I_{15}} \qquad (2)$$

Where:

- *FHP:* peak hour factor.
- I_h: intensity over the whole hour (veh/h).
- $_{15}$: vehicle count during the period of the heaviest 5 min of the hour (veh/h).

(veh).

The SATURATION GRADE relates the vehicle volume to the road capacity by the following equation:

$$x = \frac{v}{c} \qquad (3)$$

Where:

- %: degree of saturation.
- v: vehicle flow corresponding to the 15 minutes peak (veh/h).
- c: vehicle capacity (veh/h).

Corresponding $x > 1$ to a saturated flow, and $x < 1$ to an unsaturated flow.

QUALITY OF SERVICE describes how a road network section or a road transport service or roadway functions or will function from the user's point of view. It can be assessed in a number of ways. These include direct observation of factors that are perceptible to users and important to them (e.g. speed or delay), conducting user surveys, monitoring complaints and compliments about infrastructure conditions. The HCM 2010 proposes such an assessment based on a multimodal approach: car mode, pedestrian mode, cycling mode and urban public transport mode. The quality of service is characterised by the so-called Service Levels (SL), which are a quantitative classification of traffic measures and parameters. The HCM establishes 6 levels of service, ranging from NSA to NS F.

SERVICE LEVELS for urban road segment (car mode):

- SERVICE LEVEL A (NS A): mainly describes free-flow operation. Vehicles are completely free to manoeuvre within the traffic flow. Control delay at the lane intersection is minimal. Travel speed exceeds 85% of the base free-flow speed, and the volume-to-capacity ratio is not greater than 1.
- SERVICE LEVEL B (NS B): Describes reasonably free operation. The ability to manoeuvre within the traffic flow is only slightly restricted, and the control delay at the boundary intersection is not significant. The travel speed is between 67% and 85% of the base free-flow speed, and the volume-to-capacity ratio is not greater than 1.
- SERVICE LEVEL C (NS C): Describes stable operation. The ability to manoeuvre and change lanes at mid-segment locations may be more restricted than at NS B. Longer queues at the boundary intersection may contribute to lower travel speeds. The travel speed is between 50% and 67% of the base free-flow speed, and the volume-to-capacity ratio is no greater than 1.
- SERVICE LEVEL D (NS D): Indicates a less stable condition where small increases in flow can cause substantial increases in delay and decreases in travel speed. This situation may be due to adverse traffic signal progression, high volume, or inappropriate timing of traffic signals at the limit intersection. The travel speed is between 40% and 50% of the base free flow speed and the volume to capacity ratio is not greater than 1.
- SERVICE LEVEL E (NS E): Characterised by unstable operation and significant delay, which may be due to some combination of adverse progression, high volume or inappropriate timing of traffic signals at the boundary intersection. The travel speed is between 30% and 40% of the base free-flow speed and the volume to capacity ratio is not greater than 1.
- SERVICE LEVEL F (SLF): Characterised by extremely low flow speeds. Congestion is likely to occur at the boundary intersection, manifested by high delays and extensive queuing. The travel speed is 30% or less of the base free-flow speed and the volume to capacity ratio is greater than 1.

Table 1

Levels of service (LOS) in car mode for urban segments.

Travel speed as a percentage of base free flow speed (%). **base free flow velocity (%)**	**NS per Volume to Capacity Ratio**	
	< 1.0	**> 1.0**
> 85	A	F
>67 - 85	B	F
>50 - 67	C	F
>40 - 50	D	F
>30 - 40	E	F

< 30	F	F

Source: Taken from Highway Capacity Manual 2010.

LEVELS OF SERVICE for signalised intersections (automobile mode):
For the case of signalised intersections, Levels of Service can characterise the entire intersection, or each intersection approach separately according to lane groups or movement groups. In this case, the control delay and the volume-to-capacity ratio are pre-determined to characterise the respective Levels of Service. It is also a measure of driver discomfort and fuel consumption.

- SERVICE LEVEL A (SLA): Describes operations with a control delay ofO s/veh or less and a volume to capacity ratio of no more than 1. This level is generally assigned when the volume to capacity ratio is low and either the displacements are exceptionally favourable or the cycle time is very short. If due to a favourable sequence, most vehicles arrive during the green indication and travel through the intersection without stopping.
- SERVICE LEVEL B (NS B): Describes journeys with control delay between 10 and 20 s/veh and a volume-to-capacity ratio no greater than 1. This level is normally assigned when the volume-to-capacity ratio is low and the sequence is very favourable or the cycle time is short. More vehicles are stopped than in the case of NSA.
- SERVICE LEVEL C (NS C): Describes journeys with control delay between 20 and 35 s/veh and a volume-to-capacity ratio of no more than 1. Individual cycle failures may occur at this level (i.e. one or more queued vehicles cannot depart due to insufficient capacity during the cycle). The number of vehicles stopping is significant, although many vehicles still pass through the intersection without stopping.
- SERVICE LEVEL D (NS D): Describes operations with control delay between 35 y 55 s/veh and a volume to capacity ratio no greater than 1. This level is generally assigned when the volume to capacity ratio is high and either succession is ineffective or the cycle time is long. Many vehicles stop and failures are noted in individual cycles.
- SERVICE LEVEL E (NS E): Describes operations with control delay between 55 and 80 s/veh and a volume-to-capacity ratio not greater than 1. This level is normally assigned when the volume-to-capacity ratio is high, the sequence is unfavourable and the cycle time is long. Single cycle failures are frequent.
- SERVICE LEVEL F (NS F): Describes trips with a control delay greater than 80 s/veh or a volume-to-capacity ratio greater than 1. This level is usually assigned when the volume-to-capacity ratio is very high, the succession is very poor and the cycle time is long. Most cycles cannot undo the queue.

Table 2

Levels of service in car mode for signalised intersections.

Delay by Control (s/veh)	**NS per Volume-to-Capacity Ratio**	
	< 1.0	> 1.0
< 10	A	F
>10 - 20	B	F
>20 - 35	C	F
>35 - 55	D	F
>55 - 80	E	F
> 80	F	F

Source: Taken from Highway Capacity Manual 2010.

The following are the CRITERIA FOR EVALUATING THE LEVEL OF VEHICULAR SERVICE, according to the HCM 2010 methodology.
VEHICLE SERVICE LEVEL, according to the HCM 2010 methodology.
GENERAL CONSIDERATIONS:
LIMIT OF ANALYSIS of the segment is defined by the carriageway on the right-hand side of the

street y for each limiting intersection.

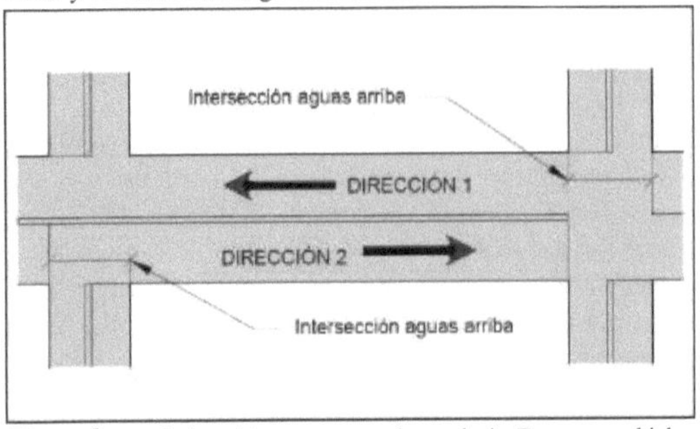

Figure 5 Boundaries of the segment under analysis. Two-way vehicle route.
Source: Taken from HCM 2010.

The LEVEL OF ANALYSIS describes the level of detail used in the application of the methodology. The following three levels of analysis are considered:

- Assessment (operational).
- Project (design).
- Planning and pre-studies.

- EVALUATION ANALYSIS; basically focused on current or short-term conditions, it requires detailed prior data with no or minimal use of default values. It is intended to provide information for decision-making on whether improvements are needed to an existing point element or section (or the entire infrastructure). It is sometimes undertaken to determine whether a wider planning study is required. It asks the question: When in the short term is an infrastructure likely to fail or cease to provide the desired level of service (to resolve this question, an estimate of the achievable service intensity under a given level of service is required).
- PROJECT ANALYSIS; uses HCM procedures to identify the characteristics required by the road network section (which should operate at the desired level of service). In some cases it uses default values for certain parameters.
- PLANNING ANALYSIS AND PREVIOUS STUDIES, with a focus on future conditions, is used when a set of alternatives is to be evaluated quickly, or when the values of certain parameters are not known and frequent use of certain parameters is required.

For the purposes of this research, the evaluation analysis is considered.

STUDY PERIOD: The methodology is based on the assumption that traffic conditions are stable during the analysis period, i.e. a systematic change over time is insignificant. For this reason, the length of the analysis period is in the range of 0.25 to 1 hour. In general, the analyst should be cautious with analysis periods exceeding 1 hour because traffic conditions are generally not stable over long periods of time and because the adverse impact of peak hours on traffic demand may not be detected in the evaluation. (Alcántara Quispe, 2018).

For analysis purposes, the URBAN STREET is separated into individual elements that are physically adjacent and operate as a single entity in service to travellers. Two elements are commonly found in an urban street system, which are points and links. A point represents the boundary between links and is represented by an intersection. A link represents a length of road between two points. A link and its boundary points are referred to as a segment (Alcántara Quispe, 2018).

The (HCM, 2010), also refers to the LIMITATIONS OF THE METHODOLOGY, in the sense that in general, the methodology described can be used to assess the performance of most traffic flows travelling along an urban street segment. However, the methodology does not address all traffic conditions or control types. The inability to replicate the influence of any one condition or control type in the methodology represents a limitation. If one or more of these limitations is believed to have a

significant influence on the performance of a specific street segment, then the analyst should consider using alternative methods or tools.
The METHODOLOGY FOR URBAN STREET SEGMENT ANALYSIS, according to the HCM 2010, considers:
- INPUT DATA: These are identified as input elements. These should be specified separately for each direction of travel in the segment and for each boundary intersection.

Table 3

Input data.

Data Category	Input elements
Traffic characteristics	Segment flow
Geometric design	Number of tracks Segment length Length of restrictive median Proportion of segments with sill Access points
Other	Length of analysis period Speed limit.

Source: Taken from Highway Capacity Manual 2010.

ACCESS POINT FLOW is defined as the number of vehicles arriving at each access point on the segment during the analysis period, divided by the duration of the analysis period. It is required for all movements intersecting at each active access point. If the segment has many access points that are considered inactive but collectively have some impact on traffic flow, these can be combined into a single equivalent active access point.
SEGMENT FLOW is defined as the number of vehicles travelling along the segment during the analysis period, divided by the duration of the analysis period. This volume is specified separately for each direction of travel along the segment. If there are one or more access points along the segment, then the flow on the segment is measured at a location between these access points (or between an access point and the boundary intersection).
SEGMENT LENGTH is measured along the street axis. If there is a difference in the two directions of travel, then an average length should be used.
The NUMBER OF ACCESS POINTS along a segment represents the number of unsignalised driveways and public streets approaching on the segment, regardless of the traffic demand entering the approach. This number is calculated separately for each side of the segment.
The PERIOD OF ANALYSIS is the time interval considered for performance evaluation. This duration is in the range of 5 minutes to 1 hour, with longer durations used for planning analysis.
LIMIT SPEED: The average travel speed is used in the methodology to evaluate segment performance. It is related to the speed limit where the speed limit reflects the environment and geometric factors that influence the driver's choice of speed. As such, the speed limit represents an input variable that can be conveniently used to estimate travel speed while limiting the need for numerous geometric and segment environment inputs.
The PARAMETERS TO BE USED IN THE EVALUATION according to (HCM, 2010), are:
- TIME IN MOTION: A procedure to determine the time in motion on the segment is described in this step. This procedure includes the calculation of the free-flow speed, an adjustment factor for the proximity between vehicles, and the additional time in motion due to sources of delay. Each calculation is explained in the following sections, culminating in the calculation of the time in motion on the segment.
- FREE FLOW SPEED: Represents the average speed of vehicles travelling through a segment under low volume conditions and without delay due to traffic control devices or other vehicles. The determination of free-flow speed is based on the calculation of the base free-flow speed and an adjustment factor for signal spacing.

- BASE FREE FLOW SPEED: Defined as the free flow speed over long segments. This includes the influence of boundary velocity, access point density, median type, presence of sill and is calculated using the following equation:

$$S_{f0} = S_0 + f_{cs} + f_A \quad (4)$$

Where:

- óy$_0$: Base free flow velocity (mi/h).
- 5$_0$: Constant speed (mi/h).
- f_{cs}: Adjustment factor for cross section (mi/h).
- f_A: Adjustment factor for access points (mi/h).

The constant velocity is determined with the following equation.

$$S_0 = 25.6 + 0.47 \cdot S_{pl} \quad (5)$$

Where:

- S_pi : velocity limit (mi/h).

Adjustment factor for cross section (mi/h):

$$f_{cs} = 1.5 \cdot p_m - 0.47 \cdot p_{curb} \cdot p_{rm} \quad (6)$$

Where:

p_{rm}: Proportion of the segment with restrictive median (decimai).
$Pcurb$ Proportion of segment with right hand side sill (decimal).
Adjustment factor for access points (mi/h):

$$f_A = -0.078 \cdot {}^{D_a}/_{N_{th}} \quad (7)$$

With

$$D_a = 5280 \cdot \left({}^{(N_{ap,s} + N_{ap,o})}/_{(L - W_i)} \right) \quad (8)$$

Where:

- D_a: Density of access points in the segment (points/mi).
- N_{th}: Number of tracks in the segment in the direction of travel.
- $N_{(aPi)\,s}$: Number of access points on the right hand side in the direction of travel.
- Nap,o : Number of access points on the right-hand side in the opposite direction of travel.
- $W_¿$: Width of upstream signalised intersection (ft).

The values for the adjustment factor for cross section and adjustment factor for access points can also be obtained taking into account the following tables:

Table 4

Adjustment factor for cross section.

Limiting velocity (mi/h)	Constant speed $S_{(o)}$ (mi/h)	Median type	Percentage with restrictive median (%)	Adjustment for cross section $f_{(rx)}$ (mi/h)	
				Without sill	With sill
25	37.4		20	0.3	-0.9
30	39.7		40	0.6	-1.4
35	42.1	Restrictive	60	0.9	-1.8
40	44.4		80	1.2	-2.2
45	46.8		100	1.5	-2.7
50	49.1	Not restrictive	Not applicable	0	-0.5
55	51.5	No median	Not applicable	0	-0.5

Source: Taken from Highway Capacity Manual 2010.

Table 5: Adjustment factor for access points.

Adjustment factor for access points.

Density of access points $D_{(a)}$ (pts/mi)	Adjustment for access points per lane, $f_{(A)}$ (mi/h)			
	1 lane	2 tracks	3 tracks	4 tracks
0	0	0	0	0
2	-0.2	-0.1	-0.1	0
4	-0.3	-0.2	-0.1	-0.1
10	-0.8	-0.4	-0.3	-0.2
20	-1.6	-0.8	-0.5	-0.4
40	-3.1	-1.6	-1	-0.8
60	-4.7	-2.3	-1.6	-1.2

Source: Taken from Highway Capacity Manual 2010.

ADJUSTMENT FOR SIGNAL SPACING: Empirical evidence suggests that a short segment length (when defined by signalised intersection boundaries) tends to influence driver choice of free-flow speed. Shorter segments have been found to have a lower free-flow speed when all other factors are held equal. The following equation is used to calculate the value of an adjustment factor that takes this influence into account.

$$f_L = 1.02 - 4.7 \cdot \frac{S_{fo} - 19.5}{max(L_S, 400)} \leq 1.0 \qquad (9)$$

Where:

- f_L: Signal spacing adjustment factor (dimensionless).
- Sf_0: Base free flow velocity (mi/h).
- L_s: Distance between adjacent signalised intersections (ft).

The FREE FLOW SPEED is calculated based on the estimates of the base free flow speed and the signal spacing adjustment factor. Alternatively, it can be entered directly by the analyst, the equation is as follows:

$$S_f = S_{fo} \cdot f_L \qquad (10)$$

Where:

- Sf: Free flow velocity (mi/h).

The VEHICLE PROXIMITY ADJUSTMENT FACTOR adjusts the free-flow moving time to account for the effect of traffic density.

The adjustments result in an increase in time in motion (and corresponding reduction in speed) with an increase in volume. The reduction in speed is a result of short intervals of separation between vehicles, associated with the high volume and the predisposition of drivers to be more cautious when intervals are short. The following equation is used to calculate the proximity adjustment factor.

$$f_V = \frac{2}{1 + \left(\frac{v_m}{52.8 \cdot N_{th} \cdot S_f}\right)^{0.21}} \qquad (11)$$

Where:

- f_v: Proximity adjustment factor.
- v_m: Demand flow ratio on the segment (veh/h).
- N_{th}: Number of lanes on the segment in the direction of travel.
- Sf. Free flow speed (mi/h).

VEHICLE TURNING DELAY: This delay is incurred by vehicles that are following turning vehicles but do not turn at the access point. For service planning and analysis the delay due to turning vehicles at access points is used by using the average segment volume. The values represent the delay due to left and right turning vehicles, the value obtained is multiplied by the number of access points in the segment to estimate the delay due to left and right turns.

Table 6: Delay due to turning vehicles.

Delay due to turning vehicles.

Average segment volume (veh/h/ln)	Vehicle delay due to number of lanes (s/veh/pt)		
	1 lane	**2 lanes**	**3 lanes**
200	0.04	0.04	0.05
300	0.08	0.08	0.09
400	0.12	0.15	0.15
500	0.18	0.25	0.15
600	0.27	0.41	0.15
700	0.39	0.72	0.15

Source: Taken from Highway Capacity Manual 2010.

The values in Table 6 represent 10% left turns and 10% right turns from the segment to the access. If the actual turn percentages are less than 10%, then the delays can be reduced proportionally. For example, if the subject access point has 5% left turns and 5% right turns, then the values should be multiplied by 0.5 (= 5/10).

DELAY CAUSED BY OTHER FACTORS: Other factors may cause a driver to slow down or incur delay while travelling along a segment. For example, a vehicle completing a parallel parking manoeuvre may cause some delay. However, this methodology only includes procedures for estimating delay due to turning vehicles.

CALCULATION OF TIME IN MOTION ON SEGMENT: Time in motion is calculated by applying the following formula:

With

$$f_x = \begin{cases} 1.00 \text{ (Movement controlled by traffic lights or stop signs)} \\ 0.00 \textit{ (Uncontrolled movement)} \\ min\backslash^{(Vth)\,/}{}_{Cth},\ 1.00] \textit{ (Yield controlled movement)} \end{cases}$$

Where:

- t_R: Time in movement in the segment.
- l_t: Time loss on departure = 2.0 if signalised, and 2.5 if stop or yield, (s).
- L: Segment length (ft).
- f_x: Adjustment factor by type of control.
- v_{th}: Demand flow ratio through the segment (veh/h).
- c_{th}: Capacity of movement through the segment (veh/h).
- $d_{(ap)\,(i)}$: Delay due to left and right turns from the street to the access point, (s/veh).
- N_{ap}: Number of influent access points approaching along the segment = N_{aps} + $p_{(a)\ p}$, $itN_{(a)\ p}$, o, (points).
- $N_{(ap)\ ,s}$: Number of nearby access points on the right-hand side of the direction of travel, (points).
- $N_{(ap)\ ,0}$: Number of access points on the opposite right hand side of the direction of travel, (points).
- *Pap.it'*- Proportion of N_{ap}that can be accessed by a left turn from the direction of travel.
- d_{other}: Delay due to other sources along the segment, (s/veh).

The other variables have been defined previously. The variables I_{1},f_x, $v_{th}$and c_{th}used in the first term of the equation apply to the motion exiting the segment at the boundary intersection.

VEHICLE PROPAGATION TIME: The delay incurred by vehicles when exiting the segment is the basis for travel time estimation. It consists of the sum of two sources of delays: delay control (due to traffic control at the intersection) and geometric delay (due to the geometry of the intersection). It is calculated by applying the following equation:

$$d_t = \frac{d_{th} \cdot v_t \cdot N_t + d_{sl} \cdot v_{sl} \cdot (1 - P_L) + d_{sr} \cdot v_{sr} \cdot (1 - P_R)}{v_{th}} \quad (13)$$

Where:

d_t: Vehicle propagation time (s/veh).

v_{th}: Vehicular demand (veh/h).

d_{th}: Lane delay (s/veh).

v_t: Lane flow (veh/h/ln).

N_t: Number of lanes (ln).

d_si : Delay of left-turning vehicles (s/veh).

- $v_{(s)\,i}$: flow rate of left-turning vehicles (veh/h).
- d_{sr}: Delay of right turning vehicles (s/veh).
- v_{sr}: flow rate of right-turning vehicles (veh/h).
- P_L: Proportion of left-turning vehicles (decimal).
- P_R: Proportion of vehicles turning right (decimal).

The TRAVEL SPEED: It is calculated by applying the following equation:

$$S_{T,seg} = \frac{3600 \cdot L}{5280 \cdot (t_R + d_t)} \quad (14)$$

Where:

- $S_{(T)\,\text{-,}\,(se)\,g}$ Vehicle travel speed through the segment (mi/h).
- L: Segment length (ft).
- t_R: Travel time (s).
- d_t: Vehicle propagation time (s/veh).

The SERVICE LEVEL: It is calculated for both directions along the segment, which is defined by two criteria: the first criterion relates to the travel speed as a percentage of the base free flow velocity and the second criterion is the relationship between the volume and the capacity of the segment (see Table N° 1).

Regarding the evaluation of the Capacity and Level of Service provided to users at a SIGNALISED INTERSECTION, the HCM 2010 for the vehicular mode proposes the use of the following parameters:

ADJUSTED SATURATION FLOW RATE:

- Saturation flow rate: .s'_o= 1900
- Group lane width factor:

Table 7

Group lane width factor.

Average Lane Width (W)	f_W
2.40 < W < 3.00	0.96
3.00 < W < 3.87	1.00
3.87 < W < 4.80	1.04

Source: Taken from Highway Capacity Manual 2010.

- Presence of heavy vehicles:

$$f_{HV} = \frac{100}{100 + P_{HV}(E_T - 1)} \quad (15)$$

Where:

f_{HV} = Factor for heavy vehicles in the group.

P_{HV} = Percentage of heavy vehicles in the group (%)

E_T = Equivalent number of cars per heavy vehicle = 2.0

- Factor for slope of access:

$$f_g = 1 - \frac{P_g}{200} \qquad (16)$$

Where:
fg = Access slope factor.
Pg= Slope at access for the lane group (%)
Positive sign if uphill and negative sign if downhill.

$$-6.00 \leq P_g \leq +10.00$$

- Factor for parking adjacent to the group:

$$f_p = \frac{N - 0.1 - \frac{18N_m}{3600}}{N} \geq 0.050 \qquad (17)$$

Where:
N_m= Parking manoeuvres adjacent to the group.
N = Number of lanes per cluster.
For parking adjacent to the cluster within 75m before reaching the intersection.
If no parking exists this value has a value of 1.00.
Maximum value for $N_m = 180\ {}^{veh}/_{h}$
Minimum value for f_p= 0.050

- Factor for blocking buses staying in the zone:

$$f_{bb} = \frac{N - \frac{14.4N_b}{3600}}{N} \geq 0.050 \qquad (18)$$

Where:
N_b = Frequency of bus stops at the access (Buses/h).
N = Number of lanes per group.
For buses stopping to pick up or drop off passengers within 75 m arriving at or passing the intersection.
Maximum value for N_b= 250 Buses/h.
Minimum value for f_{bb} = 0.050.

- Factor by type of area (f_a):

If the location is in the CBD (Central Business District) or city centre. Areas where the geometric design and traffic or pedestrian flows are such that waiting times are significantly increased: f_a = 0-90-
Location in any other area: f_a = 1-00.

- Lane utilisation adjustment factor (f_{LU}):

$$f_{LU} = \frac{v_g}{N_e v_{g1}} \qquad (19)$$

Where:
Vg = Demand for the movement group (Veh/h).
v_{gl} = Demand for the exclusive lane with the highest volume of the movement group (Veh/h).
N_e = Number of exclusive lanes in the movement group.
Used when the movement group has more than one exclusive lane.

- Left turn factor (f_{LT}):

$$f_{LT} = \frac{1}{E_L} \qquad (20)$$

Table 8

Factor per left turn in a group (f_{LT}).

Lane Type	E_L ADE Equivalent for left-turning vehicles	f_{LT}
Left turns with	**protected or unopposed phase**	
Exclusive or shared lane	1.05	0.95
Two or more ex. or shared lanes	1.09	0.92
One-way streets or T-intersections		
One exclusive or shared lane	1.18	0.85
Two or more exclusive or shared lanes	1.33	0.75

Source: Taken from Highway Capacity Manual 2010.

Table 9

Left turn factor in a group (f_{LT}): with shared lane and presence of opposing flow.

Left turns shared lane with presence of opposing flow							
Opposing flow	1	200	400	600	800	1000	1200
E_L	1.4	1.7	2.1	2.5	3.1	3.7	4.5
Left turns exclusive lane with presence of opposing flow							
Opposing flow	1	200	400	600	800	1000	1200
E_L	1.3	1.6	1.9	2.3	2.8	3.3	4.0

Source: Taken from Highway Capacity Manual 2010.

- Right turn factor in a group (f_{RT}):

$$f_{RT} = \frac{1}{E_R} \qquad (21)$$

Table 10

Right turn factor in a group (f_{RT}).

TypeofCarri[1]	ADE equivalent for right-turn vehicles	RTD
One exclusive or shared lane	1.18	0.85
Two or more exclusive or shared lanes	1.33	0.75

Source: Taken from Highway Capacity Manual 2010.

- Pedestrian and bicycle one-way roads:

$$v_{pedg} = v_{ped}\frac{C}{g_{ged}} \leq 5000 \qquad (22)$$

$$\text{Si } v_{pedg} \leq 1000\, p/h, \quad OCC_{pedg} = \frac{v_{pedg}}{2000} \qquad (23)$$

$$\text{Si } v_{pedg} > 1000\, p/h, \quad OCC_{pedg} = 0.4 + \frac{v_{pedg}}{10000} \leq 0.90 \qquad (24)$$

Where:

v_{pedg} = Pedestrian demand during green time (p/h).

v_{ped} = Pedestrian demand during peak hour (p/h).

= CTraffic light cycle length (s).

Sped= Green time for pedestrian crossing.

OCC_{pedg} = Pedestrian occupancy on green.

If there is no traffic light for pedestrians, $g_{ped}$is considered as the effective green of the phase.

Pedestrians only:

$$OCC_y = \frac{g_{ped}}{g} OCC_{pedg} \quad (25)$$

Where:
9ped = Green time for pedestrian crossing.
g = Green time for the group phase.
$OCC_{ped}g$ = Pedestrian occupancy on green.
OCCy = Major conflict zone.

- Presence of pedestrians and bicycles on two-way roads:

$$g_q = g_p - g_u \quad (26)$$

$$\text{Si } g_q < g_{ped}, \qquad OCC_{pedu} = OCC_{pedg}\left(1 - 0.5 \cdot {g_q}/{g_{ped}}\right) \quad (27)$$

$$\text{Si } g_q \geq g_{ped}, \qquad OCC_{pedu} = 0.0 \quad (28)$$

Where:
g_q = Opposite access green time (s).
g_p = Allowed green time of the access (s).
g_u = Left turn green time (s).
C = Traffic light cycle length (s).
g_{ped} = Pedestrian crossing green time (s).
OCC_{pedg} = Pedestrian occupancy on green.
OCC_{pedu} = Pedestrian occupancy after the opposite row is cleared.
Pedestrians only (for two-way roads):

$$OCC_y = \frac{g_{ped} - g_q}{g_p - g_q} OCC_{pedu} e^{-5v_o/3600} \quad (29)$$

Where:
g_q = Opposing access green time (s).
g_p = Green time allowed for the access (s).
9ped = Green time for pedestrian crossing (s).
v_0 = Opposing flow (Veh/h).
OCC_{pedu} = Pedestrian occupancy after the opposite row is cleared.
OCCy = Major conflict zone.
For receiving lanes equal to or less than turn lanes:

$$A_{pbT} = 1 - OCC_y \quad (30)$$

For receiving lanes greater than turn lanes:

$$A_{pbT} = 1 - 0.6 \cdot OCC_y \quad (31)$$

Apbı= Unoccupied zone time (s).

- Factor for pedestrians and bicycles for left-turn groups:

Table 11
Pedestrian and bicycle factor for left-turn groups ($f_{Lpb)}$.

Lane type	*f1nh*
No pedestrians or cyclists in conflict.	1.00
One-way street.	*ApbT*
Two-way street with left turn permitted.	*ApbT*
Two-way street with protected left turn permitted.	*ApbT*
Two-way street with protected left turn with no opposing flow.	1.00

Source: Taken from Highway Capacity Manual 2010.

- Pedestrian and bicycle factor for right-turn groups (f_{Rpb}) :

Table 12

Pedestrian and bicycle factor for right-turn groups (f_{Rpb}).

Lane type	tttnb
No pedestrians or cyclists in conflict.	1.00
Right turn allowed.	$ApbT$
Permitted and protected right turn.	$ApbT$
Protected right turn.	1.00

Source: Taken from Highway Capacity Manual 2010.

Determination of the PROPORTION OF VEHICLES COMING IN GREEN.

$$P = R_P \frac{g}{C} \qquad (32)$$

Where:

Rp = Platoon flow.

P = Proportion of vehicles arriving during green (in decimal).

∂= Effective green time of the access (s).

C = Traffic light cycle length (s).

Table 13

Platoon flow for progress characterisation.

Platoon flow	Arrival type	Progress
0.33	1	Very poor
0.67	2	Unfavourable
1.00	3	Random arrivals
1.33	4	Favourable
1.67	5	Very favourable
2.00	6	Exceptionally favourable

Source: Taken from Highway Capacity Manual 2010.

CAPACITY AND DEGREE OF SATURATION.

- Adjusted saturation flow for each lane of each lane group.

$$s = s_0 f_w f_{HV} f_g f_p f_{bb} f_a f_{LU} f_{LT} f_{RT} f_{Lpb} f_{Rpb} \qquad (33)$$

Where:

s: Adjusted saturation flow rate (veh/h/ln).

5_0: Base saturation flow rate (pc/h/ln).

f_w: Adjustment factor for lane width.

f_{HV}: Adjustment factor for heavy vehicles in traffic flow.

fg : Adjustment factor for degree of approach.

f_p: Adjustment factor for the existence of a parking lane and parking activity adjacent to the lane group. f_p: Adjustment factor for the existence of a parking lane and parking activity adjacent to the lane group.

f_{bb}: Adjustment factor for the blocking effect of local buses stopping within the intersection area.

f_a: Adjustment factor for type of area.

f_{LU}: Adjustment factor for lane usage.

f_{LT}: Adjustment factor for the presence of a vehicle on the left in a group of lanes.

f_{RT}: Adjustment factor for the presence of the vehicle on the right in a lane group.

f_{Lpb}: Pedestrian adjustment factor for left-turn groups.

f_{Rpb}: Pedestrian-bicycle adjustment factor for right-turn groups.

Capacity and volume-capacity ratio (degree of saturation):

$$Q = c = Ns\frac{g}{C} \qquad (34)$$

$$X = \frac{v}{c} = \frac{q}{Q} \qquad (35)$$

Where:
Q = c = Capacity (ADE/h).
N = Number of lanes.
5 = Adjusted saturation flow rate (ADE/h).
g = Effective group green time (s).
C = Traffic light cycle length (s).
q = V = Adjusted demand (ADE/h).
X = Degree of Saturation.

DELAYS AT SIGNALISED INTERSECTIONS.

The delay calculated in this step represents the average control delay experienced by all arriving vehicles during the analysis period. It includes any delay incurred by these vehicles that are still queued after the end of the analysis period. The control delay for a given lane group is calculated using Eq:

$$d = d_1 + d_2 + d_3 \qquad (36)$$

Where:
d: Control delay (s/veh).
d^. Uniform delay (s/veh).
d_2: Incremental delay (s/veh).
d_3: Initial queue delay (s/veh).

- Uniform delay (d_1).

At a signalised intersection, the first type of delay is known as uniform delay (d_1). It is so called because demand is assumed to arrive at the intersection at a uniform or constant rate q.
uniform or constant rate q.

$$d_1 = \frac{0.5C\left(1-\frac{g}{C}\right)^2}{1-\left[min(1,X)\frac{g}{C}\right]} \qquad (37)$$

Where:
$d_\pm$: Uniform delay (s/veh).
C: Traffic light cycle (s).
g : Green time.
'Y: Degree of saturation.

- Incremental delay (d_2)

$$d_2 = 900T\left[(X-1)+\sqrt{(X-1)^2+\frac{8KIX}{cT}}\right] \qquad (38)$$

Where:
T: Length of analysis period (0.25 hours).
K: Incremental delay factor. K= 0.5 for preset intersections.
I = Adjustment factor for upstream intersection inputs. $I=1$ for isolated intersections.
c= Lane group capacity.

For the calculation of REPRESENTATIVE DELAY AND SERVICE LEVEL at the entire intersection, the following weighted average shall be performed:

$$D_i = \frac{\sum(d_a \cdot V_i)}{\sum V_i} \qquad (39)$$

Where:

Di =Delay at the signalised intersection (s/veh).

Vi =Hourly vehicle volume at an intersection approach (veh/h).

d_a=Average delay at the access (s/veh).

Some DEFINITIONS OF BASIC TERMS:

AFFOROUS: Vehicle capacity is the number of vehicles or persons passing through a given section of road.

GULLY: A surface drainage work that is usually constructed transversally to the axis of the road or following the orientation of the watercourse. Culverts can be made of wood, stone, concrete, metal and others.

VEHICLE DENSITY: The number of vehicles occupying a stretch of given length averaged over the length in units of vehicles per kilometre.

TRAFFIC CONTROL DEVICES: Signs, markings, traffic lights and auxiliary devices that have the function of facilitating the driver's strict observance of the rules governing vehicular traffic, both on roads and on city streets.

ROAD ELEMENTS: The set of physical components of the road, such as road surface, berms, ditches, works of art, drainage, road safety elements and environment.

ROAD INFRASTRUCTURE: Any road, artery, street or railway, including its complementary works, of rural or urban character of public and private domain and use.

TRAFFIC INTENSITY: The number of vehicles passing through a given profile during a time interval of less than one hour, usually 15 minutes. Maximum traffic intensity: The number of vehicles passing a given profile during the 15 minutes of maximum vehicle demand.

VEHICLE VOLUME: The number of vehicles passing a given point or cross-section of a given lane or carriageway during a given period.

1.2. **Problem formulation**

What is the capacity and levels of service of Mario Urteaga Avenue, section: Jr. Dos de Mayo to El Inca Oval, according to the HCM 2010 methodology, Cajamarca 2018?

1.3. **Objectives**

1.3.1. General objective

To analyse the capacity and levels of service of Av. Mario Urteaga, section: Jr.
Dos de Mayo to Óvalo el Inca; according to the methodology HCM 2010, Cajamarca 2018.

Specific objectives

- Determine the capacity of the Mario Urteaga Avenue, section: Jr. Dos de Mayo to Óvalo el Inca: considered as a road segment; according to the HCM 2010 methodology.
- Determine the level of service of Av. Mario Urteaga, section: Jr. Dos de Mayo to Óvalo el Inca: considered as a road segment; according to HCM 2010 methodology.
- Determine the average delay of the traffic-light intersection: Av. Mario Urteaga - Jr. Urrelo; according to HCM 2010 methodology.
- Determine the levels of service of all accesses, and as a whole of the traffic signalised intersection: Av. Mario Urteaga - Jr. Urrelo; according to HCM 2010 methodology.

1.4. **Hypotheses**

1.4.1. General hypothesis

The capacity and levels of service of Av. Mario Urteaga, section: Jr. Dos de Mayo to Óvalo el Inca, correspond to the most critical levels according to the HCM 2010 classification.

CHAPTER II

METHODOLOGY

2.1. Type of investigation

It is applied, as it has practical objectives. Cross-sectional prospective observational, because there is no intervention by the researcher, the data are planned and measured only on one occasion.

2.2. Population and sample

The population is made up of all the roads in the city of Cajamarca that fall within the classification of collector roads.

2.3. Materials, instruments and methods

LOCATION OF THE STUDY SECTION:

A. POLITICAL LOCATION:

Country	:	Peru
Region :	:	Cajamarca
Department :	:	Cajamarca
Province :	:	Cajamarca
District :	:	Cajamarca

B. GEOGRAPHICAL LOCATION:

- U.T.M. coordinates (WGS-84).
- The study is located in zone 17M.

> Section: Mario Urteaga Avenue from Óvalo el Inca to Jr. Dos de Mayo.

- Starting Point

North	: 9208004.53	Latitude	: -7.158322658966728
East	: 774923.01	Longitude	: -78.51082129911394
Coordinate :	: 2701.61	Height	: 2701.61

- End Point

North	: 9208395.10	Latitude	: -7.1548139180843755
East	: 774494.28	Longitude	: -78.51471967999227
Height	: 2711.05	Height	: 2711.05

- The point where the survey was carried out has the following coordinates:

North	: 9208081.22	Latitude	: -7.157634080027542
East	: 774831.22	Longitude	: -78.51165559782788
Height	: 2702.68	Height	: 2702.68

Figure 6 Location of the study site.
Source: Google Earth, 2018

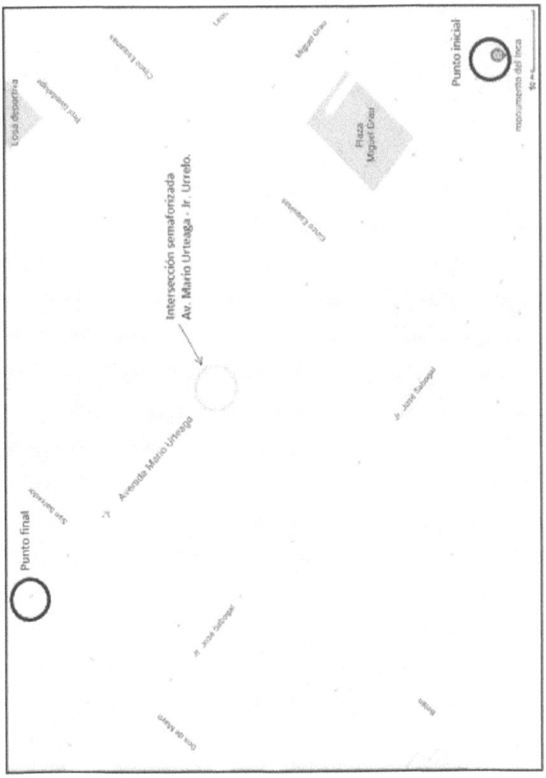

Figure 7 Location of the study site.

Source: Google Maps 2018.

The methodology used for the determination of the Capacity and Level of Service of the indicated urban road was developed based on the Highway Capacity Manual 2010, a publication of the Transportation Research Board, a United States Institute whose mission is to promote innovation and progress in transportation through research.

2.4. Techniques and instruments for data collection and analysis

Direct observation was applied as a technique for data collection, with the use of video surveillance cameras strategically located at the study points; and the following formats were used for the systematisation of information, both for the study segment and for the intersection considered during one week of study in each case:

Table 14

Type of vehicles circulating on the road under study.

Vehículo Tipo	Detalle
L5	
M1	
N1	
M2	
M2	
M3	

Source: (MTC, Vehicle Classification
Vehicle Classification and Standardisation of Vehicle

Vehicle Recordable Characteristics. , 2006)

Instrument 1

Vehicle flow rate. Direction:

I. General data							
Thesis:	ANALYSIS OF THE CAPACITY AND LEVELS OF SERVICE OF AVENIDA MARIO URTEAGA, TRAMO: JR. DOS DE MAYO TO EL INCA OVAL; ACCORDING TO THE METHODOLOGY HCM 2010, CAJAMARCA 2018.						
Thesis thesis:	EVER ROJAS HUAMÁN						
City:	CAJAMARCA						
Street:	AV. MARIO URTEAGA 191				[redacted]		
File:	VEHICLE FLOW RATE. SHIFT:						
Date: Start time:					End time:		
II. Data	DIRECTION:						
TIME:	TRIMOT L5	CAR M1	TRUCKS PICKUP	RURAL Combi M2	MICRO M2	BUS M3	TOTAL
	fj						

I. Datos generales		
Tesis:	ANÁLISIS DE LA CAPACIDAD Y NIVELES DE SERVICIO DE LA AVENIDA MARIO URTEAGA, TRAMO: JR. DOS DE MAYO HASTA EL ÓVALO EL INCA; SEGÚN LA METODOLOGÍA HCM 2010, CAJAMARCA 2018.	
Tesista:	EVER ROJAS HUAMÁN	
Ciudad:	CAJAMARCA	
Calle:	AV. MARIO URTEAGA 191	
Ficha:	TASA DE FLUJO VEHICULAR. TURNO:	
Fecha:	Hora de inicio:	Hora de término:

II. Datos	SENTIDO:						
HORARIO	TRIMOTO L5	AUTO M1	CAMIONETAS		MICRO M2	BUS M3	TOTAL
			PICKUP	RURAL Combi M2			

Instrument 2

Vehicle capacity. Traffic light intersection: Av. Mario Urteaga - Jr. Urrelo.

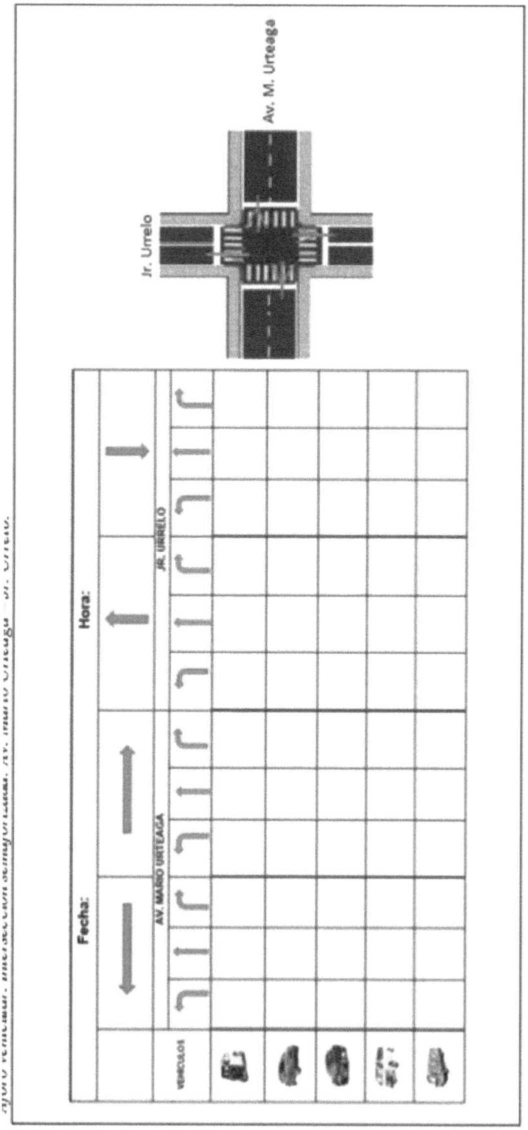

Instrument 3

Vehicle capacity. Segment of road: Av. Mario Urteaga.

Fecha:		Hora:
VEHICULOS	AV. MARIO URTEAGA	
	↑	↓

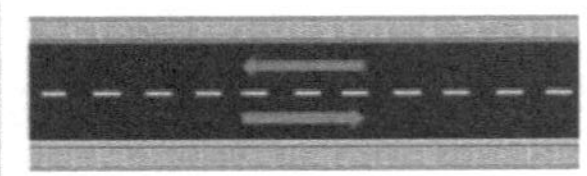

Av. Mario Urteaga

Instrument 4

Speed calculation. Segment of road: Mario Urteaga Avenue - Jr. Urrelo.

1. General Data					
Thesis:	ANALYSIS OF THE CAPACITY AND LEVELS OF SERVICE OF MARIO URTEAGA AVENUE, SECTION: JR. DOS DE MAYO TO EL INCA OVAL; ACCORDING TO THE METHODOLOGY HCM 2010, CAJAMARCA 2OIS.				
Thesis thesis:	EVER ROJAS HUAMÁN				
City:	CAJAMARCA				
Street:	AV. MARIO URTEAGA 191				
Tramo:	JR. DOS DE MAYO - OVALO EL INCA				
File:	CALCULATION OF SPEEDS.				
Date:	Dell 05/11/1S to 09/11/1S	Start time:	10:00 hrs	Finish time:	12:00 hrs

II. Data

PROJECT VEHICLE		Time {seconds}					Average weekly time	Distance {m}	Speed {m/s)	(km/h)
		Mon	Tue	Wed	Thu Thur	Fri				
JR. DOS DE MAYO - EL INCA OVAL										
	1°									
	2"									
	3"									
	4°									
	5"									
	6"									
	7°									
	8"									
	9°									
	10"									
Average daily time (seconds):										

2.5. Procedure

The information was collected through a vehicle count with the support of video surveillance cameras strategically located in the segments and intersections; using a counting format by type of vehicle every 15 minutes from 6:30 am to 9:00 pm during one week.

The following elements were considered for the study:

- The segment with vehicular circulation from Óvalo el Inca to Jr. Dos de Mayo and from Jr. Dos de Mayo to Óvalo el Inca.
- Intersection (traffic light) of Mario Urteaga Ave. with Urrelo Jr.

In order to calculate the CAPACITY AND SERVICE LEVEL of the segment considered, the following steps were applied

The following general steps were applied for the calculation of the CAPACITY AND SERVICE LEVEL of the segment under consideration:

Step 1: Determine the input elements

Step 2: Determine the time in motion

Step 3: Determine the vehicular propagation time.

Step 4: Determine the travel speed

Step 5: Determine Level of Service (LOS)

Figure 8 Steps for the assessment of the vehicular level of service of a road section applying the HCM 2010 methodology.

Step 1, - INPUT ELEMENTS.

Table 15 Segment characteristics.

Segment characteristics.

Description	Track segment
Segment length:	540m
Number of lanes:	2
Number of accesses:	13
Average carriageway width:	llm
Pumping:	2%

Presence of ditches:	yes
Presence of culverts:	yes
Presence of concrete manhole covers:	yes
Width of ditches:	0.40, 0.30

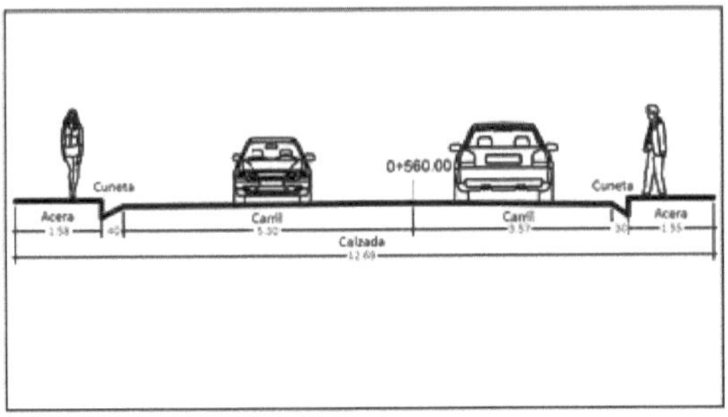

Figure 9 Typical cross-section of the road studied.

Description

Segment length:
Number of lanes:
Number of accesses:
Average carriageway width:
Pumping:
Presence of ditches:
Presence of culverts:
Presence of concrete manhole covers:
Width of ditches:

- Vehicle capacity (maximum vehicle intensity), of the road segment:

$$I_{máx} = \frac{\textit{Número de vehículos}}{\textit{Tiempo (hrs)}} = \frac{383}{0.25} = 1532\ {}^{veh}/_{h}$$

Here, Number of vehicles, corresponds to the vehicle count during the period of the5 min most loaded Peak Demand Hour (see Figure 21).

- Peak Hour Factor (PHF) of the track segment (Equation 2):

$$FHP = \frac{I_h}{4 \cdot I_{15}} = \frac{1446}{4 \cdot 383} = 0.94$$

The value of $I_{(h)}$ (intensity in the whole hour), corresponds to the VHMD (see Figure 15).

- Degree of saturation. Using Equation 3, and with the previous result of the

Vehicle Capacity:

$$x = \frac{v}{c} = \frac{\frac{383\ veh}{15\ min} \cdot \left(\frac{60\ min}{1\ h}\right)}{1532} = 1.00$$

Step 2,- TIME IN MOTION:

Table 16

Input elements for the segment.

Data category	Input elements	Segment
Segment characteristics	Segment capacity (veh/h)	1532
traffic	Demand flow in the exclusive group, $v_{(t)}$ (veh/h)	1446

Geometric design	Traffic signalised intersection width (ft)	36.09
	Number of lanes in exclusive lane group (N_t)	2
	Segment length (ft)	1771.65
	Number of right side accesses ($N_{(ap)(>s)}$)	5
	Number of accesses on left side ($N_{(ap)(>0)}$)	8
	Proportion of segment with restrictive median (decimal)	0
	Proportion of segment with right-hand side sill	1
Other	Speed limit $S_{(p)i}$, (mi/h)	37.30
	Length of analysis period, T (h)	0.25

Note: According to the National Traffic Regulation Art. 162, the speed limit for avenues in urban areas is 60 km/h (37.3 mi/h) (MTC, 2009).

- Density of access points (Equation 8):

$$D_a = 5280 \cdot \frac{N_{ap,s} + N_{ap,o}}{L - W_i} = 5280 \cdot \frac{5+8}{1771.65 - 36.09} = 39.55$$

- Adjustment factor for access points (Equation 7):

$$f_A = -0.078 \cdot \frac{D_a}{N_{th}} = -0.078 \cdot \frac{39.55}{1} = -3.08 \ {}^{mi}/_{h}$$

- Adjustment factor for cross section (Equation (6)):

$$f_{cs} = 1.5 \cdot p_{rm} - 0.47 \cdot p_{curb} - 3.7 \cdot p_{curb} \cdot p_{rm} = 1.5 \cdot 0 - 0.47 \cdot 1 - 3.7 \cdot 1 \cdot 0 = -0.47$$

- Constant velocity (Equation (5)):

$$S_0 = 25.6 + 0.47 \cdot S_{pl} = 25.6 + 0.47 \cdot 37.30 = 43.13 \ {}^{mi}/_{h}$$

- Base free flow velocity (Equation 4):

$$S_{fo} = S_0 + f_{cs} + f_A = 43.13 + (-0.47) + (-3.08) = 39.58 \ {}^{mi}/_{h}$$

- Adjustment for signal spacing (Equation 9):

$$f_L = 1.02 - 4.7 \cdot \frac{S_{fo} - 19.5}{max(L_s, 400)} = 1.02 - 4.7 \cdot \frac{39.58 - 19.5}{400} = 0.78 \leq 1.0$$

Free-flow speed (Equation 10):

$$S_f = S_{fo} \cdot f_L = 39.58 \cdot 0.78 = 30.87 \ {}^{mi}/_{h}$$

Vehicle proximity factor (Equation 11):

$$f_v = \frac{2}{1 + \left(1 - \frac{v_m}{52.8 \cdot N_{th} \cdot S_f}\right)^{0.21}} = \frac{2}{1 + \left(1 - \frac{1446}{52.8 \cdot 1 \cdot 30.87}\right)^{0.21}} = 1.22$$

Time in motion in the segment (Equation 12):

$$t_R = \frac{6.0 - l_1}{0.0025 \cdot L} f_x + \frac{3600 \cdot L}{5280 \cdot S_f} f_v + \sum_{i=1}^{N_{ap}} d_{ap,i} + d_{other}$$

$$t_R = \frac{6.0 - 2}{0.0025 \cdot 1771.65} \cdot 1 + \frac{3600 \cdot 1771.65}{5280 \cdot 30.87} \cdot 1.22 + 5 \cdot 0.03 + 8 \cdot 0.03 = 49.03 \ s$$

Step 3.- VEHICULAR PROPAGATION TIME.

- Vehicle propagation time (Equation 13 and Table 16):

$$d_t = \frac{d_{th} v_t N_t + d_{sl} v_{sl}(1 - P_L) + d_{sr} v_{sr}(1 - P_R)}{v_{th}}$$

$$d_t = \frac{25.20 \cdot 1446 \cdot 2 + 3 \cdot 145 \cdot (1 - 0.10) + 2 \cdot 145 \cdot (1 - 0.10)}{1446} = 50.9\ {}^{s}/_{veh}$$

Step 4.- TRAVEL SPEED.

- Travel speed (Equation 14):

$$S_{T,seg} = \frac{3600 \cdot L}{5280 \cdot (t_R + d_t)} = \frac{3600 \cdot 1771.65}{5280 \cdot (49.03 + 50.9)} = 12.08\ mi/h$$

Step 5,- SERVICE LEVEL (NS).

Percentage ratio between the displacement velocity and the base flow velocity; in addition to the velocity-to-capacity ratio:

$$\% = \frac{S_t}{S_{fo}}\% = \frac{12.08}{39.58} x100\% = 30.52\%; \qquad \frac{v}{c} = 0.94$$

and the use of Table 1, results in a Level of Service E.

- Calculation of the congestion index (Equation 1):

$$CI = \frac{T_r}{T_l} = \frac{16.10}{12.38} = 1.30$$

For the process of calculating the LEVEL OF SERVICE ASSOCIATED WITH THE SEMAFORIZED INTERSECTION: Av. Mario Urteaga - Jr. Urrelo; the following general steps were carried out:

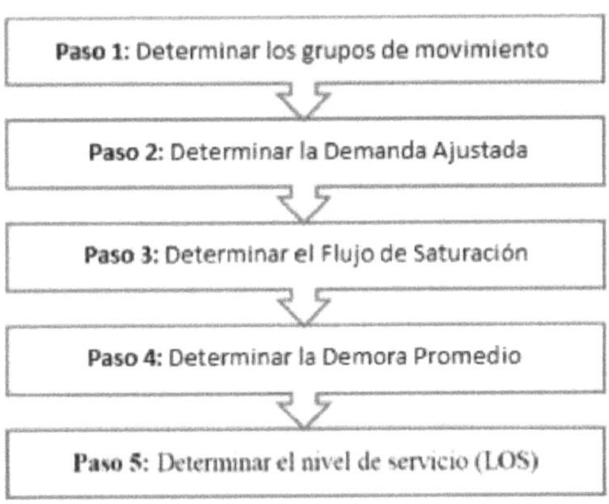

Figure 10 Steps for the evaluation of the level of vehicular service of traffic signalised intersection, applying the HCM 2010 methodology.

The determination of the Hourly Volume of Maximum Demand (VHMD), for each access movement, was carried out from the data in Annex 43 - 54.

Pasol.-MOVEMENTGROUPS.

The definition of the Movement Groups is shown in Figure 11.

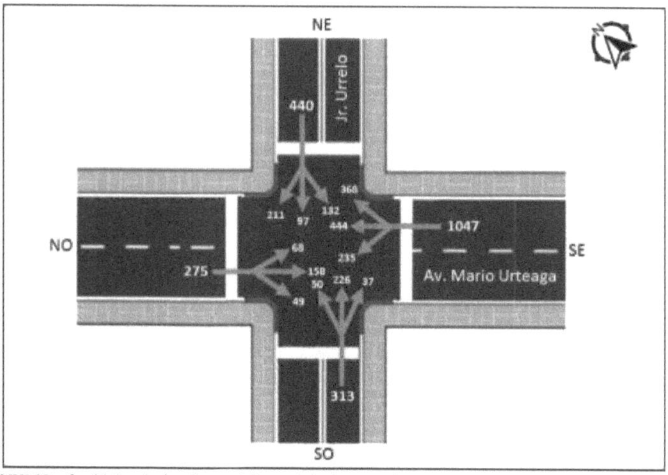

Figure 11 VHMD of vehicles during the peak hour at the intersection.

Table 17

Data for traffic signalised intersection movement groups:
Av. Mario Urteaga - Jr. Urrelo.

Access	**GM**	**tv(m.)**	Pfcv(%)	*ET*	*Pt(%~)*	**N** N_m	N_b	**Zone**
SE	1	4.84	0	2.5	1	1 2	0	centre
NO	2	4.84	0	2.5	-1	1 1	0	centre
SO	3	4.62	0	2.5	1	1 1	0	centre
NE	4	4.62	0	2.5	-1	1 1	0	centre

Step 2.- ADJUSTED DEMAND.

- Hourly factor of maximum demand (FHMD); movement: Left turn. Direction: Av. Mario Urteaga - Jr. Urrelo (SE - NO). (See Annex 43)

$$\text{FHMD} = \frac{Q}{4 \cdot Q_{15}} = \frac{235}{4 \cdot 63} = 0.93$$

Peak Hourly Demand Factor (PODF); movement: Straight through. Mario Urteaga Av. Mario Urteaga. Direction: SE - NW. (See Annex 44)

$$\text{FHMD} = \frac{Q}{4 \cdot Q_{15}} = \frac{444}{4 \cdot 119} = 0.93$$

Peak Hourly Factor of Maximum Demand (HFMD); movement: Right turn. Direction: Av. Mario Urteaga - Jr. Urrelo (SE - NO). (See Annex 45)

$$\text{FHMD} = \frac{Q}{4 \cdot Q_{15}} = \frac{368}{4 \cdot 102} = 0.90$$

Hourly factor of maximum demand (FHMD); movement: left turn. Direction: Av. Mario Urteaga - Jr. Urrelo (NW - SE) (See Annex 46).

$$\text{FHMD} = \frac{Q}{4 \cdot Q_{15}} = \frac{68}{4 \cdot 19} = 0.89$$

Peak Hourly Factor of Maximum Demand (PHMD); movement: Straight through. Mario Urteaga Av. Mario Urteaga. Direction: NW - SE. (See Annex 47).

$$FHMD = \frac{Q}{4 \cdot Q_{15}} = \frac{158}{4 \cdot 48} = 0.82$$

Hourly factor of maximum demand (FHMD); movement: Right turn. Direction: Av. Mario Urteaga - Jr. Urrelo (NW - SE) (See Annex 48).

$$FHMD = \frac{Q}{4 \cdot Q_{15}} = \frac{49}{4 \cdot 15} = 0.82$$

Hourly factor of maximum demand (FHMD); movement: left turn. Direction: Jr. Urrelo - Av. Mario Urteaga (SW - NE). (See Annex 49)

$$FHMD = \frac{Q}{4 \cdot Q_{15}} = \frac{50}{4 \cdot 20} = 0.62$$

Peak Hourly Demand Factor (PODF); movement: Straight through. Jr. Urrelo. Direction: SW - NE (See Annex 50).

$$FHMD = \frac{Q}{4 \cdot Q_{15}} = \frac{226}{4 \cdot 76} = 0.74$$

Peak Hourly Demand Factor (PMDF); movement: Right turn. Direction: Jr. Urrelo - Av. Mario Urteaga (SW - NE). (See Annex 51)

$$FHMD = \frac{Q}{4 \cdot Q_{15}} = \frac{37}{4 \cdot 17} = 0.54$$

Hourly factor of maximum demand (FHMD); movement: left turn. Direction: Jr. Urrelo - Av. Mario Urteaga (NE - SW) (See Annex 52).

$$FHMD = \frac{Q}{4 \cdot Q_{15}} = \frac{132}{4 \cdot 46} = 0.72$$

Peak Hourly Demand Factor (PHMD); movement: Straight through. Jr. Urrelo. Direction: NE - SW. (See Annex 53)

$$FHMD = \frac{Q}{4 \cdot Q_{15}} = \frac{97}{4 \cdot 28} = 0.87$$

Hourly factor of maximum demand (FHMD); movement: Right turn. Direction: Jr. Urrelo - Av. Mario Urteaga (NE - SW) (See Annex 54).

$$FHMD = \frac{Q}{4 \cdot Q_{15}} = \frac{211}{4 \cdot 64} = 0.8$$

Table 18: Initial conditions at intersection: Av.

Initial conditions at intersection: Av.
Mario Urteaga - Jr. Urrelo.

Elements	Quantity
Accesses	4
Lanes	4
Movements	12
Track groups	4
Movement groups	4

- Equivalent vehicles grouped by movement: Left turn. Direction: Av. Mario Urteaga - Jr. Urrelo (SE - NW).

$$q_{mov} = \frac{VHMD}{FHMD \cdot f_{hv}} = \frac{235}{0.93 \cdot 1} = 253$$

Note: Table 42 summarises the total equivalent vehicle calculations for the 12 movement types for the study intersection.

Table 19: Adjusted demand at the study intersection

Adjusted demand at the signalised intersection: Av. Mario Urteaga - Jr. Urrelo.

Phase	Access	Lane	Mov	VHMD	FHMD	f_{hv}	Яmoг	ЯcacazzИ	Яacccезo	Яccc	9GM
Φ1	SE	1		368	0.90	1	409	■ 1139	1139	1139	1139
			F	444	0.93	1	477				
			I	235	0.93	1	253				
	NO	2	D	49	0.82	1	60	■ 329	329	329	329
			F	158	0.82	1	193				
			I	68	0.89	1	76				
φ2	SO	3	D	37	0.54	1	69	■ 455	455	455	455
			F	226	0.74	1	305				
			I	50	0.62	1	81				
	NE	4	D	211	0.82	1	257	■ 551	551	551	551
			F	97	0.87	1	111				
			I	132	0.72	1	183				

Step 3,- SATURATION FLOW.

PARAMETERS FOR LEFT TURNS; Intersection Av. Mario Urteaga - Jr. Urrelo:

Table 20

Data for left turn factor f_{LT}. Traffic light intersection: Av. Mario Urteaga - Jr. Urrelo.

Access	GM	Case	Flow Op.	Interp. E_L
SE	1	3	158	1.616
NO	2	3	444	2.188
SO	3	3	97	1.494
NE	4	3	226	1.752

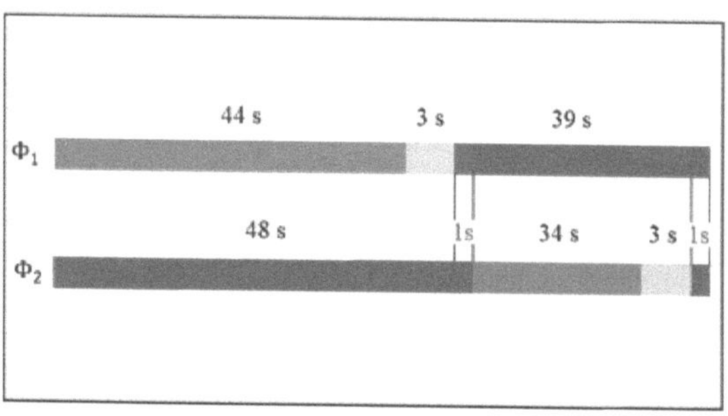

Figure 12. Traffic light phases: Oí (for Av. Mario Urteaga) y $_{\Phi(2)}$ (for Jr. Urrelo).

- Pedestrian volume in green time for SE access by left turn, (Equation 22):

$$v_{pedg} = \frac{v_{ped} \cdot C}{g_{ped}} = \frac{62 \cdot 86}{44} = 121.18$$

- Pedestrian occupancy at green time, SE access by left turn, (Equation 23):

$$OCC_{pedg} = \frac{v_{pedg}}{2000} = \frac{121.18}{2000} = 0.06$$

- Green time access opposite SE access (Equation 26):

$$g_q = g_p - g_u = 44 - 44 = 0$$

- Pedestrian occupancy after the opposite row is cleared, SE access (Equation 27):

$$OCC_{pedu} = OCC_{pedg}\left(1 - 0.5\frac{g_q}{g_{ped}}\right) = 0.06 \cdot \left(1 - 0.5\frac{0}{44}\right) = 0.06$$

Major conflict zone, SE access (Equation 29):

$$OCC_y = \frac{g_{ped} - g_q}{g_p - g_q} OCC_{pedu} e^{\frac{-5v_0}{3600}} = \frac{44 - 0}{44 - 0} \cdot 0.06 \cdot e^{\frac{-5 \cdot 158}{3600}} = 0.05$$

- Time zone unoccupied (s); SE access (Equation 30):

$$A_{pbT} = 1 - OCC_y = 1 - 0.05 = 0.95$$

Table 34 summarises the data and parameters, with respect to all accesses, for the pedestrian and bicycle left turn factor f_{Lpb}.

RIGHT TURN PARAMETERS: Intersection Av. Mario Urteaga - Jr.

Urrelo:

Table 21

Data for right turn factor f_{RT}. Traffic light intersection: Av. Mario Urteaga - Jr. Urrelo.

Access	Movement Groups (GM)	Lanes
SE NW SW NE	1	1
	2	1
	3	1
	4	1

- Pedestrian volume in green time for SE right-turn access,

(Equation 22):

$$v_{pedg} = \frac{v_{ped} \cdot C}{g_{ped}} = \frac{70 \cdot 86}{44} = 136.82$$

- Pedestrian occupancy at green time; SE right-turn access (Equation 23):

$$OCC_{pedg} = \frac{v_{pedg}}{2000} = \frac{136.82}{2000} = 0.07$$

Table 43 summarises the data and parameters, with respect to all accesses, for the right turn factor for pedestrian and bicycle presence $f_{(Rp)b}$.

Step 3,- SATURATION FLOW.

- Factor for slope at access; SE access (Equation 16):

$$f_g = 1 - \frac{P_g}{200} = 1 - \frac{1}{200} = 1.00$$

- Factor for parking adjacent to the cluster; SE access ((Equation 17)):

$$f_p = \frac{N - 0.1 - \frac{18N_m}{3600}}{N} = \frac{1 - 0.1 - \frac{18 \cdot 1}{3600}}{1} = 0.90$$

- Factor for left turns (f_{LT}); SE access (Equation 20):

$$f_{LT} = \frac{1}{E_L} = \frac{1}{1.62} = 0.62$$

- Factor for right turn into a cluster (f_{RT}); SE access (Equation 21):

$$f_{RT} = \frac{1}{E_R} = \frac{1}{1.18} = 0.85$$

- Saturation flow s; SE access (Equation 33):

$$s = 1900 \cdot 1 \cdot 1 \cdot 1 \cdot 0.9 \cdot 1 \cdot 0.9 \cdot 1 \cdot 0.62 \cdot 0.85 \cdot 0.95 \cdot 0.96 = 728$$

Table 36 summarises the data and parameters, with respect to all accesses, for the calculation of the saturation flow.

- Supply *Q;* SE access (Equation 34):

$$Q = \frac{s \cdot N \cdot V_{ai}}{C} = \frac{728 \cdot 1 \cdot 44}{86} = 372$$

- Degree of saturation for each SE access (Equation 35).

$$X = \frac{q}{Q} = 3.06$$

Table 46 summarises the data and parameters, for all accesses, for the calculation of the degree of saturation.

Steps 4 and 5.- AVERAGE DELAY AND SERVICE LEVELS.

- AVERAGE DELAY of the SE access of the traffic-light intersection: Av. Mario Urteaga - Jr. Urteaga - Jr. Urrelo (Equation 36, 37, 38):

$$d_1 = \frac{0.5 \cdot C \cdot \left(1 - \frac{g}{C}\right)^2}{1 - \left[min(1, X) \cdot \frac{g}{C}\right]} = \frac{0.5 \cdot 86 \cdot \left(1 - \frac{44}{86}\right)^2}{1 - \left[1 \cdot \frac{44}{86}\right]} = 21\, seg$$

$$d_2 = 900 \cdot T \cdot \left[(X-1) + \sqrt{(X-1)^2 + \frac{8 \cdot k \cdot I \cdot X}{c \cdot T}}\right] =$$

$$= 900 \cdot 0.25 \cdot \left[(3.06 - 1) + \sqrt{(3.06-1)^2 + \frac{8 \cdot 0.5 \cdot 1 \cdot 3.06}{1136 \cdot 0.25}}\right] = 929\, seg$$

$$D_t = d_1 + d_2 + d_3 = 21 + 803 + 0 = 950\, seg$$

According to Table 2, the Level of Service is F.

- AVERAGE DELAY of the NW access of the signalised intersection: Av. Mario Urteaga - Jr. Urteaga - Jr. Urrelo.

$$d_1 = \frac{0.5 \cdot C \cdot \left(1 - \frac{g}{C}\right)^2}{1 - \left[min(1, X) \cdot \frac{g}{C}\right]} = \frac{0.5 \cdot 86 \cdot \left(1 - \frac{44}{86}\right)^2}{1 - \left[1 \cdot \frac{44}{86}\right]} = 21\ seg$$

$$d_2 = 900 \cdot T \cdot \left[(X-1) + \sqrt{(X-1)^2 + \frac{8 \cdot k \cdot I \cdot X}{c \cdot T}}\right] =$$

$$= 900 \cdot 0.25 \cdot \left[(1.18-1) + \sqrt{(1.18-1)^2 + \frac{8 \cdot 0.5 \cdot 1 \cdot 1.18}{328 \cdot 0.25}}\right] = 108\ seg$$

$$D_t = d_1 + d_2 + d_3 = 21 + 108 + 0 = 129\ seg$$

According to Table 2, the Level of Service is F.

- AVERAGE DELAY of the SW access to the signalised intersection: Mario Urteaga Av. - Jr. Urrelo.

$$d_1 = \frac{0.5 \cdot C \cdot \left(1 - \frac{g}{C}\right)^2}{1 - \left[min(1, X) \cdot \frac{g}{C}\right]} = \frac{0.5 \cdot 86 \cdot \left(1 - \frac{34}{86}\right)^2}{1 - \left[1 \cdot \frac{34}{86}\right]} = 26\ seg$$

$$d_2 = 900 \cdot T \cdot \left[(X-1) + \sqrt{(X-1)^2 + \frac{8 \cdot k \cdot I \cdot X}{c \cdot T}}\right] =$$

$$= 900 \cdot 0.25 \cdot \left[(1.52-1) + \sqrt{(1.52-1)^2 + \frac{8 \cdot 0.5 \cdot 1 \cdot 1.52}{452 \cdot 0.25}}\right] = 245\ seg$$

$$D_t = d_1 + d_2 + d_3 = 26 + 245 + 0 = 271\ seg$$

According to Table 2, the Level of Service is F.

- AVERAGE DELAY for the NE approach to the traffic-light intersection: Av. Mario Urteaga - Jr. Urteaga - Jr. Urrelo.

$$d_1 = \frac{0.5 \cdot C \cdot \left(1 - \frac{g}{C}\right)^2}{1 - \left[min(1, X) \cdot \frac{g}{C}\right]} = \frac{0.5 \cdot 86 \cdot \left(1 - \frac{34}{86}\right)^2}{1 - \left[1 \cdot \frac{34}{86}\right]} = 26\ seg$$

$$d_2 = 900 \cdot T \cdot \left[(X-1) + \sqrt{(X-1)^2 + \frac{8 \cdot k \cdot I \cdot X}{c \cdot T}}\right] =$$

$$= 900 \cdot 0.25 \cdot \left[(2.14 - 1) + \sqrt{(2.14 - 1)^2 + \frac{8 \cdot 0.5 \cdot 1 \cdot 2.14}{552 \cdot 0.25}}\right] = 519\, seg$$

$$D_t = d_1 + PF \cdot d_2 + d_3 = 26 + 1 \cdot 519 + 0 = 545\, seg$$

According to Table 2, the Level of Service is F.
DELAY AND LEVEL OF SERVICE FOR THE ENTIRE SIGNALISED INTERSECTION:
Av. Mario Urteaga - Ir. Urrelo: (Equation 39)

$$D_i = \frac{\sum(d_a \cdot V_i)}{\sum V_i} = \frac{950 \cdot 1047 + 129 \cdot 275 + 271 \cdot 313 + 545 \cdot 440}{1047 + 275 + 313 + 440} = 653\, s$$

According to Table 2, the Level of Service is F.

CHAPTER III

RESULTS

- TABLES AND FIGURES OF RESULTS OF THE ROAD SEGMENT ANALYSIS.

Table 23

Total mixed vehicle composition per day of the studied road segment (from 05 to 11 November 2018).

Weekday	Total vehicles	%
Monday	17260	16.85
Tuesday	16650	16.25
Wednesday	17218	16.81
Thursday	16736	16.34
Friday	17996	17.57
Saturday	10338	10.09
Sunday	6250	6.09
TOTAL:	102448	100

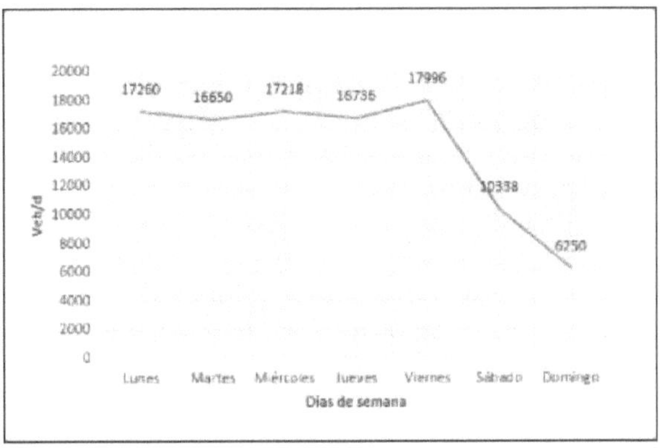

Figure 13. Mixed vehicle composition per day, of the studied road section (from 05 to 11 November 2018).

November 2018).

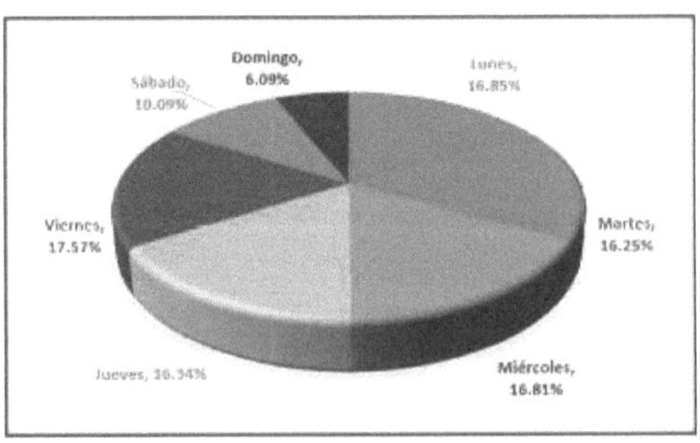

Figure 14. Percentage composition of mixed vehicles per day, of the studied road section (from 5 to 11 November 2019).

11 November 2019).

Table 24

Total weekly vehicle composition of the studied road section.

Type of vehicle	Number	%
Trimoto:	59692	58.3
Car:	19192	18.7
Pick up:	8753	8.5
Combi:	13528	13.2
Micro:	1283	1.3
Bus:	0	0
TOTAL:	102448	100

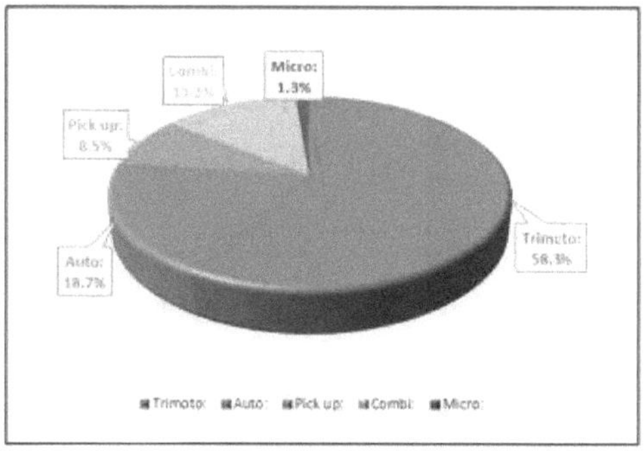

Figure 15. Weekly percentage share by vehicle type, on the studied road section; Monday 05 to Sunday 11 November 2018.

Table 25

Weekly composition of vehicular traffic. Direction: Óvalo El Inca - Jr. Dos de Mayo.

Type of vehicle	Number	%
Trimoto:	32628	49.8
Car:	14802	22.6
Pick up:	6741	10.3
Combi:	10183	15.5
Micro:	1112	1.8
Bus:	0	0
TOTAL:	65466	100

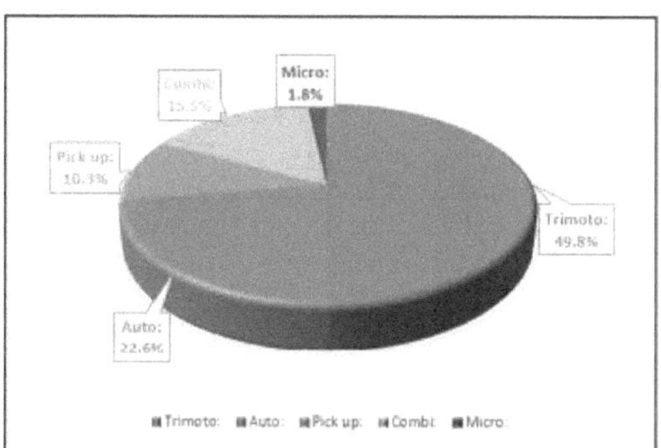

Figure 16. Percentage participation by type of vehicle on the stretch of road studied; from Monday 05 to Sunday 11,

November 2018. Direction: Óvalo El Inca - Jr. Dos de Mayo.

Table 26

Weekly composition of vehicle traffic. Direction: Jr. Dos de Mayo - Óvalo El Inca.

Type of vehicle	Number	%
Trimoto:	27064	73.2
Car:	4390	11.9
Pick up:	2012	5.4
Combi:	3345	9.0
Micro:	171	0.5
Bus:	0	0
TOTAL:	36982	100

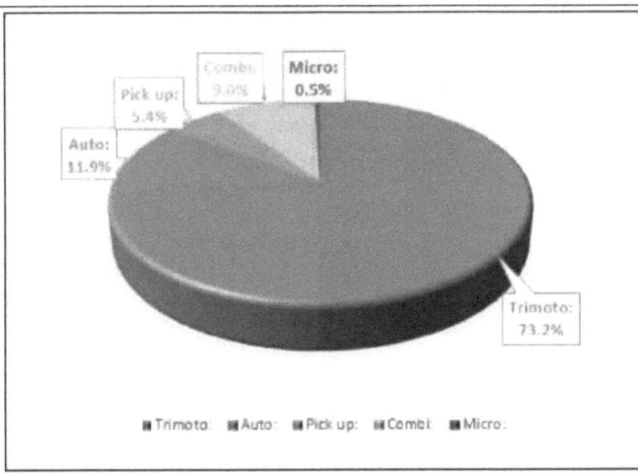

Figure 17 Percentage participation by type of vehicle on the stretch of road studied; from Monday 05 to Sunday 11 November 2018. Direction: Jr. Dos de Mayo - Óvalo El Inca.

Table 27

Summary of weekly mixed vehicle traffic flow in both directions: El Inca Oval - Jr. Dos de Mayo (SE - NW); Jr. Dos de Mayo - El Inca Oval (NW - SE).

Day	SE -NO	NO-SE
Monday	11185	6075
Tuesday	11038	5612
Wednesday	11005	6213
Thursday	10890	5846
Friday	11229	6767
Saturday	6707	3631
Sunday	3412	2838
TOTAL:	65466	36982

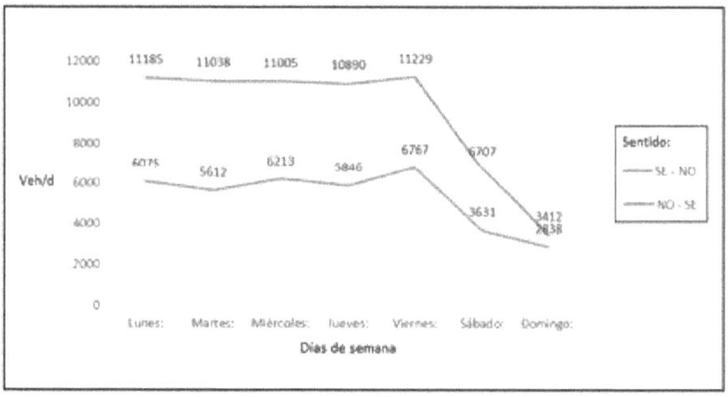

Figura 18 Comparative table of weekly mixed traffic flow for both directions: El Inca Oval - Jr. Dos de Mayo (SE - NW); Jr. Dos de Mayo - El Inca Oval (NW - SE).

Table 28

Summary of vehicle flow for the section of road under study; Monday 05 November 2018.

Time	Total	Cumulative Sum
6:30-6:45	149	
6:45 - 7:00	[redacted]	
7:00-7:15	[redacted]	
7:15-7:30	440	1344
7:30 - 7:45	[redacted]	[redacted]
7:45 - 8:00	298	1527
8:00-8:15	310	1404
8:15 - 8:30	341	1305
8:30-8:45	320	1269
8:45-9:00	297	1268
9:00-9:15	254	1212
9:15-9:30	291	1162
9:30-9:45	313	1155
9:45 - 10:00	310	1168
10:00- 10:15	310	1224
10:15 - 10:30	291	1224
10:30- 10:45	293	1204
10:45 - 11:00	313	1207
11:00- 11:15	290	1187
11:15 - 11:30	334	1230
11:30- 11:45	283	1220
11:45 - 12:00	277	1184
12:00- 12:15	309	1203
12:15 - 12:30	296	1165
12:30- 12:45	317	1199
12:45 - 13:00	292	1214
13:00- 13:15	341	1246
13:15 - 13:30	296	1246
13:30- 13:45	306	1235
13:45 - 14:00	240	1183
14:00- 14:15	238	1080
14:15 - 14:30	271	1055
14:30- 14:45	285	1034
14:45 - 15:00	265	1059
15:00- 15:15	303	1124
15:15 - 15:30	305	1158
15:30- 15:45	295	1168
15:45 - 16:00	282	1185
16:00- 16:15	324	1206
16:15 - 16:30	288	1189
16:30- 16:45	320	1214
16:45 - 17:00	304	1236
17:00- 17:15	318	1230
17:15 - 17:30	299	1241
6:30-6:45	307	1228
6:45 - 7:00	319	1243
7:00-7:15	317	1242
7:15-7:30	336	1279
7:30 - 7:45	302	1274

7:45 - 8:00	293	1248
8:00-8:15	300	1231
8:15 - 8:30	250	1145
8:30-8:45	297	1140
8:45-9:00	321	1168
9:00-9:15	285	1153
9:15-9:30	199	1102

Table 29

Summary of the vehicle capacity of the section of road under study; Tuesday 06 November 2018.

Time	Total	Cumulative Sum
6:30-6:45	135	
6:45-7:00	306	
7:00-7:15	454	
7:15-7:30	411	1306
7:30-7:45	313	1484
7:45-8:00	302	1480
8:00-8:15	323	1349
8:15-8:30	314	1252
8:30-8:45	352	1291
8:45-9:00	291	1280
9:00-9:15	271	1228
9:15-9:30	293	1207
9:30-9:45	312	1167
9:45 - 10:00	312	1188
10:00- 10:15	310	1227
10:15 - 10:30	300	1234
10:30- 10:45	284	1206
10:45 - 11:00	309	1203
11:00- 11:15	301	1194
11:15 - 11:30	289	1183
11:30- 11:45	281	1180
11:45 - 12:00	263	1134
12:00- 12:15	296	1129
12:15 - 12:30	241	1081
12:30- 12:45	301	1101
12:45 - 13:00	293	1131
13:00- 13:15	315	1150
13:15 - 13:30	284	1193
13:30- 13:45	284	1176
13:45 - 14:00	255	1138
14:00- 14:15	219	1042
14:15 - 14:30	249	1007
14:30- 14:45	276	999
14:45 - 15:00	259	1003
15:00- 15:15	325	1109
15:15 - 15:30	276	1136
15:30- 15:45	316	1176
15:45 - 16:00	293	1210
16:00- 16:15	310	1195
16:15 - 16:30	287	1206
16:30- 16:45	319	1209
16:45 - 17:00	305	1221
17:00- 17:15	306	1217
17:15 - 17:30	303	1233
6:30-6:45	289	1203
6:45-7:00	297	1195
7:00-7:15	280	1169
7:15-7:30	276	1142
7:30-7:45	256	1109
7:45-8:00	243	1055
8:00-8:15	272	1047
8:15-8:30	274	1045
8:30-8:45	287	1076
8:45-9:00	276	1109
9:00-9:15	270	1107
9:15-9:30	208	1041

Table 30

Summary of the vehicle capacity of the section of road under study; Wednesday 07 November 2018.

Time	Total	Cumulative Sum
6:30-6:45	136	

6:45 - 7:00	316	
7:00-7:15	503	
7:15-7:30	[redacted]	1408
7:30 - 7:45	[redacted]	[redacted]
7:45 - 8:00	298	1565
8:00-8:15	308	1370
8:15 - 8:30	293	1210
8:30-8:45	332	1231
8:45-9:00	296	1229
9:00-9:15	249	1170
9:15-9:30	288	1165
9:30-9:45	286	1119
9:45 - 10:00	302	1125
10:00- 10:15	292	1168
10:15 - 10:30	310	1190
10:30- 10:45	280	1184
10:45 - 11:00	316	1198
11:00- 11:15	286	1192
11:15 - 11:30	301	1183
11:30- 11:45	276	1179
11:45 - 12:00	273	1136
12:00- 12:15	311	1161
12:15 - 12:30	290	1150
12:30- 12:45	286	1160
12:45 - 13:00	301	1188
13:00- 13:15	329	1206
13:15 - 13:30	334	1250
13:30- 13:45	316	1280
13:45 - 14:00	294	1273
14:00- 14:15	283	1227
14:15 - 14:30	290	1183
14:30- 14:45	286	1153
14:45 - 15:00	242	1101
15:00- 15:15	334	1152
15:15 - 15:30	277	1139
15:30- 15:45	314	1167
15:45 - 16:00	303	1228
16:00- 16:15	303	1197
16:15 - 16:30	280	1200
16:30- 16:45	288	1174
16:45 - 17:00	293	1164
17:00- 17:15	334	1195
17:15 - 17:30	301	1216
6:30-6:45	295	1223
6:45 - 7:00	311	1241
7:00-7:15	335	1242
7:15-7:30	281	1222
7:30 - 7:45	264	1191
7:45 - 8:00	275	1155
8:00-8:15	287	1107
8:15 - 8:30	254	1080
8:30-8:45	312	1128
8:45-9:00	290	1143
9:00-9:15	262	1118
9:15-9:30	247	lili

Table 31

Summary of the vehicle capacity of the section of road under study; Thursday 08 November 2018.

Time	Total	Cumulative sum
6:30-6:45	154	
6:45-7:00	345	
7:00-7:15	561	
7:15-7:30	[redacted]	-
7:30-7:45	[redacted]	[redacted]
7:45-8:00	218	1507
8:00-8:15	287	1233
8:15-8:30	276	1114
8:30-8:45	277	1058
8:45-9:00	287	1127

9:00-9:15	280	1120
9:15-9:30	289	1133
9:30-9:45	305	1161
9:45 - 10:00	310	1184
10:00- 10:15	308	1212
10:15 - 10:30	307	1230
10:30- 10:45	298	1223
10:45 - 11:00	320	1233
11:00- 11:15	296	1221
11:15 - 11:30	344	1258
11:30- 11:45	250	1210
11:45 - 12:00	288	1178
12:00- 12:15	299	1181
12:15 - 12:30	264	1101
12:30- 12:45	242	1093
12:45 - 13:00	234	1039
13:00- 13:15	263	1003
13:15 - 13:30	254	993
13:30- 13:45	291	1042
13:45 - 14:00	215	1023
14:00- 14:15	262	1022
14:15 - 14:30	267	1035
14:30- 14:45	265	1009
14:45 - 15:00	244	1038
15:00- 15:15	343	1119
15:15 - 15:30	291	1143
15:30- 15:45	321	1199
15:45 - 16:00	280	1235
16:00- 16:15	318	1210
16:15 - 16:30	301	1220
16:30- 16:45	300	1199
16:45 - 17:00	296	1215
17:00- 17:15	316	1213
17:15 - 17:30	275	1187
6:30-6:45	301	1188
6:45-7:00	284	1176
7:00-7:15	316	1176
7:15-7:30	291	1192
7:30-7:45	285	1176
7:45-8:00	298	1190
8:00-8:15	297	1171
8:15-8:30	304	1184
8:30-8:45	352	1251
8:45-9:00	288	1241
9:00-9:15	236	1180
9:15-9:30	218	1094

Table 32

Summary of the vehicle capacity of the section of road under study; Friday 09 November 2018.

Time	Total	Cumulative Sum
6:30-6:45	109	
6:45 - 7:00	342	
7:00-7:15	369	
7:15-7:30	311	1131
7:30 - 7:45	378	1400
7:45 - 8:00	383	1441
8:00-8:15	314	1386
8:15 - 8:30	371	1446
8:30-8:45	341	1409
8:45-9:00	372	1398
9:00-9:15	279	1363
9:15-9:30	278	1270
9:30-9:45	309	1238
9:45 - 10:00	313	1179
10:00- 10:15	278	1178
10:15 - 10:30	312	1212
10:30- 10:45	291	1194
10:45 - 11:00	309	1190
11:00- 11:15	305	1217
11:15 - 11:30	385	1290
11:30- 11:45	340	1339
11:45 - 12:00	334	1364

12:00- 12:15	324	1383
12:15 - 12:30	349	1347
12:30- 12:45	345	1352
12:45 - 13:00	320	1338
13:00- 13:15	280	1294
13:15 - 13:30	312	1257
13:30- 13:45	316	1228
13:45 - 14:00	257	1165
14:00- 14:15	256	1141
14:15 - 14:30	280	1109
14:30- 14:45	275	1068
14:45 - 15:00	256	1067
15:00- 15:15	319	1130
15:15 - 15:30	286	1136
15:30- 15:45	320	1181
15:45 - 16:00	308	1233
16:00- 16:15	332	1246
16:15 - 16:30	300	1260
16:30- 16:45	318	1258
16:45 - 17:00	306	1256
17:00- 17:15	330	1254
17:15 - 17:30	299	1253
6:30-6:45	301	1236
6:45 - 7:00	341	1271
7:00-7:15	275	1216
7:15-7:30	362	1279
7:30 - 7:45	324	1302
7:45 - 8:00	354	1315
8:00-8:15	312	1352
8:15 - 8:30	288	1278
8:30-8:45	379	1333
8:45-9:00	407	1386
9:00-9:15	269	1343
9:15-9:30	228	1283

Table 33

Summary of the vehicle capacity of the section of road under study; Saturday 10 November 2018.

Time	Total	Cumulative Sum
6:30-6:45	110	
6:45-7:00	183	
7:00-7:15	168	
7:15-7:30	162	623
7:30-7:45	201	714
7:45-8:00	189	720
8:00-8:15	192	744
8:15-8:30	176	758
8:30-8:45	220	777
8:45-9:00	150	738
9:00-9:15	201	747
9:15-9:30	209	780
9:30-9:45	143	703
9:45 - 10:00	169	722
10:00- 10:15	160	681
10:15 - 10:30	164	636
10:30- 10:45	161	654
10:45 - 11:00	197	682
11:00- 11:15	157	679
11:15 - 11:30	173	688
11:30- 11:45	166	693
11:45 - 12:00	161	657
12:00- 12:15	150	650
12:15 - 12:30	167	644
12:30- 12:45	163	641
12:45 - 13:00	193	673
13:00- 13:15	186	709
13:15 - 13:30	159	701
13:30- 13:45	196	734
13:45 - 14:00	180	721
14:00- 14:15	199	734
14:15 - 14:30	208	783
14:30- 14:45	149	736
14:45 - 15:00	153	709

15:00- 15:15	213	723
15:15 - 15:30	210	725
15:30- 15:45	151	727
15:45 - 16:00	182	756
16:00- 16:15	204	747
16:15 - 16:30	205	742
16:30- 16:45	183	774
16:45 - 17:00	185	777
17:00- 17:15	193	766
17:15 - 17:30	156	717
6:30-6:45	227	761
6:45-7:00	159	735
7:00-7:15	213	755
7:15-7:30	191	790
7:30-7:45	166	729
7:45-8:00	167	737
8:00-8:15	181	705
8:15-8:30	188	702
8:30-8:45	186	722
8:45-9:00	164	719
9:00-9:15	182	720
9:15-9:30	145	677

Table 34

Summary of vehicle capacity for the section of road under study; Sunday 11 November 2018.

Time	Total	Cumulative Sum
6:30 -6:45	114	
6:45 - 7:00	134	
7:00 -7:15	113	
7:15 -7:30	101	462
7:30 - 7:45	98	446
7:45 - 8:00	100	412
8:00 -8:15	106	405
8:15 - 8:30	127	431
8:30 -8:45	107	440
8:45 - 9:00	120	460
9:00 - 9:15	97	451
9:15 - 9:30	121	445
9:30 - 9:45	107	445
9:45 - 10:00	114	439
10:00 - 10:15	112	454
10:15 - 10:30	114	447
10:30 - 10:45	108	448
10:45 - 11:00	107	441
11:00 - 11:15	92	421
11:15 - 11:30	105	412
11:30 - 11:45	110	414
11:45 - 12:00	111	418
12:00 - 12:15	115	441
12:15 - 12:30	93	429
12:30 - 12:45	116	435
12:45 - 13:00	99	423
13:00 - 13:15	90	398
13:15 - 13:30	114	419
13:30 - 13:45	105	408
13:45 - 14:00	93	402
14:00 - 14:15	110	422
14:15 - 14:30	89	397
14:30 - 14:45	118	410
14:45 - 15:00	115	432
15:00 - 15:15	99	421
15:15 - 15:30	122	454
15:30 - 15:45	107	443
15:45 - 16:00	103	431
16:00 - 16:15	82	414
16:15 - 16:30	99	391
16:30 - 16:45	102	386
16:45 - 17:00	110	393
17:00 - 17:15	104	415
17:15 - 17:30	95	411
6:30 -6:45	123	432
6:45 - 7:00	116	438

7:00 -7:15	118	452
7:15 -7:30	109	466
7:30 - 7:45	123	466
7:45 - 8:00	91	441
8:00 -8:15	94	417
8:15 - 8:30	138	446
8:30 -8:45	107	430
8:45 -9:00	99	438
9:00 -9:15	130	474
9:15 -9:30	100	436

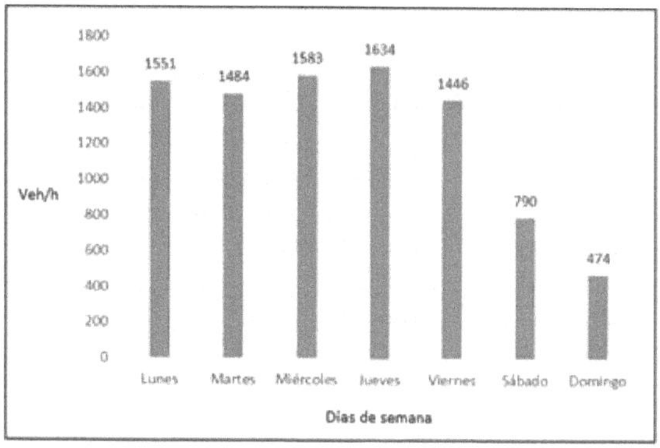

Figura 19 Comparative table of maximum hourly vehicle demand on the section of road studied, during the week of Monday 05 to Sunday 11 November 2018.

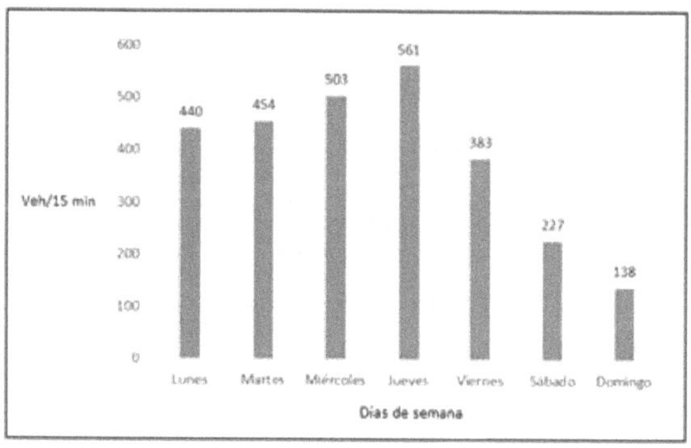

Figura 20 Comparative table of maximum vehicle flows (15 min interval) with respect to the hour of maximum demand per day, during the week of Monday 05 to Sunday 11 November 2018.

Due to the fact that the section under study has the highest vehicle demand on Fridays, the hour of maximum vehicle volume and the 15 minutes of maximum demand on that day were determined.

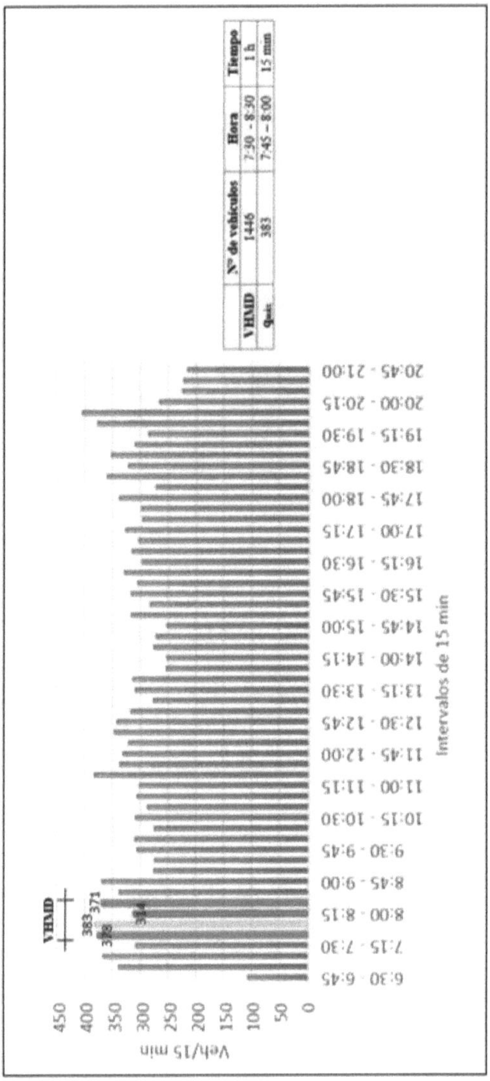

Figure 21. VHMD, $q_m$áx, and vehicle flow on Friday 09 November 2018.

Table 35

Traffic parameters for the road segment.

Parameter	Segment
Peak hour vehicle volume:	1446
Vehicle volume corresponding to the 15 min. peak demand:	383
Vehicle capacity (v/h):	1532
Vehicle demand flow:	1446
Peak hour factor:	0.94
Saturation degree:	1.00

Table 36

Values of base free flow velocity, constant velocity, adjustment factor for cross section and access points, for the road segment.

Parámetros	Segmento
$L\ (ft)$	1771.65
$N_{ap,s}$	5
$N_{ap,o}$	8
$Wi_{ap,s}$	36.09
N_{th}	2
$S_{pl}\ (^{mi}/_{h})$	37.30
p_{rm}	0
p_{curb}	1
$Da\ (^{pto}/_{mi})$	39.55
$f_A\ (^{mi}/_{h})$	-3.08
$f_{cs}\ (^{mi}/_{h})$	-0.47
$S_{f0}\ (^{mi}/_{h})$	39.58
$So\ (^{mi}/_{h})$	43.13

Table 37

Time-in-motion values obtained for the track segment.

Parameters	Segment
Parámetros	**Segmento**
f_L	0.78
$S_f\ (^{mi}/_{h})$	30.87
v_m	1446
f_v	1.22
l_1	2
$d_{ap,i}$	0.01
f_x	1
$t_R\ (s)$	49.03

Table 38

Vehicle propagation time for the track segment.

Parameters	Segment

Parámetros	Segmento
$d_{th}\ (s/veh)$	25.2
$v_t\ (veh/h/ln)$	1446
N_t	2
$d_{sl}\ (s/veh)$	3
$v_{sl}\ (veh/h)$	145
P_L	0.10
$d_{sr}\ (s/veh)$	2
$v_{sr}\ (veh/h)$	145
P_L	0.10
$v_{th}\ (veh/h)$	1446
$d_t\ (s/veh)$	50.90

Table 39

Travelling speed on the track segment.

Parámetros	Valores
$L\ (ft)$	1771.65
$d_t\ (s/veh)$	50.90
$t_R(s/veh)$	49.03
$S_t\ (mi/h)$	12.08

Table 40

Percentage ratio of travel speed to base flow velocity, and volume to capacity ratio.

Parámetros	Valores
$S_t\ (mi/h)$	12.08
$S_{f0}\ (mi/h)$	39.58
%	30.52
$Volumen/capacidad$	0.94

> The track segment has level of service E.

Table 41

Congestion Index.

Parameters	Values
Travel time	16.10
Free flow time	12.38
Congestion Index	1.30

- TABLES OF RESULTS OF THE SIGNALISED INTERSECTION ANALYSIS.

Table 42

Summary of the maximum hourly demand factors (FHMD) and equivalent vehicles (tfmov) for each movement on the approaches to the traffic-light intersection: Av. Mario Urteaga - Jr. Urrelo.

Access movements	FHMD	4mov
Left turn. Direction: Av. Mario Urteaga - Jr. Urrelo (SE - NW).	0.93	253
Direct Pass. Mario Urteaga Av. Direction: SE - NO.	0.93	477
Right turn. Direction: Av. Mario Urteaga - Jr. Urrelo (SE - NO).	0.90	409
Turn left. Direction: Av. Mario Urteaga - Jr. Urrelo (NW - SE).	0.89	76
Direct Pass. Mario Urteaga Av. Direction: NW - SE.	0.82	193
Right turn. Heading: Av. Mario Urteaga - Jr. Urrelo (NW - SE).	0.82	60
Turn left. Direction: Jr. Urrelo - Av. Mario Urteaga (SW - NE).	0.62	81
Direct Pass. Jr. Urrelo. Direction: SW - NE.	0.74	305
Right turn. Direction: Jr. Urrelo - Av. Mario Urteaga (SW - NE).	0.54	69
Turn left. Direction: Jr. Urrelo - Av. Mario Urteaga (NE - SW).	0.72	183
Direct Pass. Jr. Urrelo. Direction: NE - SW.	0.87	111
Right turn. Direction: Jr. Urrelo - Av. Mario Urteaga (NE - SW).	0.82	257

Table 43

Summary of data and parameters, for all accesses, for the left turn factor for pedestrian and bicycle presence f_{Lpb}. Traffic-light intersection: Av. Mario Urteaga - Jr. Urrelo.

Acc	GM	C	Sped	Va 'ped	VA 'pedg	OCC_{pedg}	Sp	Su	9_4	OCC_{pedu}	v_0	occ_y	Carr. r.	Carr. v.	^pbT
SE	1	86	44	62	121.18	0.06	44	44	0	0.06	158	0.05	1	1	0.95
NO	2	86	44	70	136.82	0.07	44	44	0	0.07	444	0.04	1	1	0.96
SO	3	86	34	110	278.24	0.14	34	34	0	0.14	97	0.12	1	1	0.88
NE	4	86	34	120	303.53	0.15	34	34	0	0.15	226	0.11	1	1	0.89

Table 44

Summary of data and parameters, for all accesses, for the pedestrian and bicycle right turn factor f_{Rpb}. Traffic-light intersection: Av. Mario Urteaga - Jr. Urrelo.

Acc	GM	C	Sped	VH 'ped	V 'pedg	H OCC_{pedg}	9	OCCy	Carr. r.	Carr. v	ApbT
SE	1	86	44	70	136.82	0.07	44	0.04	1	1	0.96
NO	2	86	44	62	121.18	0.06	44	0.05	1	1	0.95
SO	3	86	34	120	303.53	0.15	34	0.11	1	1	0.89
NE	4	86	34	110	278.24	0.14	34	0.12	1	1	0.88

Table 45

Calculation of saturation flow for each access. Traffic-light intersection: Av. Mario Urteaga - Jr. Urteaga - Jr. Urrelo.

Phase	Access	GM	s_o	fw	IHV		fp	fbb	f_a	fiu	fur	I_{RT}	flpb	IRpb	5
Φ1	SE	1	1900	1	1	1.00	0.90	1	0.9	1	0.62	0.85	0.95	0.96	728
	NO	2	1900	1	1	1.01	0.90	1	0.9	1	0.46	0.85	0.96	0.95	543
Φ(2)	SO	3	1900	1	1	1.00	1	1	0.9	1	0.67	0.85	0.88	0.89	756
	NE	4	1900	1	1	1.01	1	1	0.9	1	0.57	0.85	0.89	0.88	651

Table 46

Calculation of the degree of saturation for each access. Traffic-light intersection: Av. Mario Urteaga - Jr. Urrelo.

Phase	Access	GM	N	Vai	C	Q	q	X
Φı	SE	1	1	44	86	372	1139	3.06
	NO	2	1	44	86	278	329	1.18
ϕ(2)	SO	3	1	34	86	299	455	1.52
	NE	4	1	34	86	257	551	2.14

Table 47 Summary of Average Delays and Levels of Service for each access and as a whole.

Summary of Average Delays and Levels of Service for each access and as a whole; Traffic Light Intersection: Av. Mario Urteaga - Jr. Urrelo.

Access	Average Delays (s)	Overall delay (s)	Level of Service (NS)
SE	950		F
NO	129	635	F F
SO	271		F
NE	545		F

CHAPTER IV

DISCUSSION AND CONCLUSIONS

4.1 Discussion

The results of the present Thesis, as in the work of Sabando Santibáñez, (2017) concludes that the proposal to determine the level of service in urban streets, based on the application of the HCM 2010, in general lines is applicable to the national reality, allowing to complement the development of urban road projects and helping to make decisions on whether the performance of a road facility is acceptable or not; however, it will always be necessary to analyse the values of the parameters, for each specific case.

In comparison with the work of Cuenca Poma (2014), although it is true that they coincided in the determination of capacity and levels of service from geometric information (length, slope, overrun, etc.); and the HCM methodology; however, the difference in terms of the methodology for counting vehicle capacity is highlighted. In Cuenca Poma's work, pneumatic vehicle counters are used in complete 24-hour cycles, in contrast to the present work in which the vehicle capacity was measured using video surveillance cameras with cycles of 14.5 hours per day; weighing the advantage of using a longer cycle in the first case, with the disadvantage of using pneumatic counters (risk of erroneous counting in the case of vehicles with more than two axles).

For the study of the traffic-light intersection, resorting to the comparative results between the HCM 2000 and HCM 2010, carried out by Agreda Ríos & Parra Alarcón (2017) in their work on traffic-light sections of San Luis Avenue in the city of Lima; we consider that in the present work there are no incompatibilities concerning the average bus blocking time per stop, for the simple reason that we have not considered such bus blocking. The analysis carried out by Agreda Ríos & Parra Alarcón based on their own experience allowed them to identify an incompatibility of 2.4% in the average bus time per stop due to the use of the

HCM methodology. This fact suggests as a recommendation to carry out a global calibration plan of the HCM methodology, with the respective adaptation of parameter values so that the existing formulas and tables respond better to the Peruvian case.

On the other hand, as in the work of Chuquihuaccha Mauricio & Galván Aldoradín (2016), it is also possible to improve the level of service at a signalised intersection by optimally modulating the existing traffic light timings. From the results of the HCM methodology for the case of intersections, we can consider whether or not it is necessary to improve the supply conditions (capacity) of intersections. The traffic light intersection of Av. Mario Urteaga - Jr. Urrelo, which is the subject of part of this study, despite having a distribution of relevant traffic light phases in the sense of having a higher proportion of green duration on Av. Mario Urtega (which has a higher vehicle volume), compared to Jr. Urrelo; however, the individual levels of service of each access and as a whole according to the HCM is F, which means that the delay per control exceeds 80 s/veh and a volume-capacity ratio greater than 80 s/veh, which means that the delay per control exceeds 80 s/veh and a volume-capacity ratio higher than 80 s/veh, which means that the delay per control exceeds 80 s/veh and a volume-capacity ratio higher than 80 s/veh.In this sense, the recommendation should be oriented towards reducing the traffic light cycle to the point of reducing or even eliminating queues in the best of cases (always maintaining the relevant proportion of green time indicated for Av. Mario Urteaga with respect to Jr. Urrelo); for which it is also necessary to control the volumes entering the segment from the accesses of influence (Jr. Ayacucho, Jr. Cinco Esquinas) that contribute with an important vehicular demand.

With the work of Arguedas Bemal & Mosqueira Grosso (2018), which manages to improve the level of service of the intersection of Avenidas Aviación and Del Aire in the city of Lima, going from a level "E" to a level "D"; from the actions of modification of the cycle of the existing traffic light and redirection of the traffic lights, it is possible to improve the level of service of the intersection of Avenidas Aviación and Del Aire in the city of Lima.

There is no major strategic comparison with the present Thesis, since heavy vehicles do not circulate on Mario Urteaga Avenue.
Alcántara Quispe (2018), in his study of the San Martín de Porres Avenue located between Atahualpa Avenue and Argentina Avenue in the city of Cajamarca, manages to obtain different results of the levels of service in both segments of the road (this is due to the fact that in one direction there is greater traffic flow than in the opposite direction). In contrast to this, the results of the present thesis do not consider the study of two separate road segments, but the study of a single road segment with consideration of the lanes in both directions, because there is no central road separator at the time of the study.
Finally, Llanos Rimarachín (2018), in his analysis of the level of service of the busiest traffic-light intersections on Av. Hoyos Rubio, also managed to improve the level of service from F to D at the most critical intersections, proposing improvements related basically to changing the traffic light cycle.

4.2 Conclusions

In accordance with the objectives set out, it was possible to determine:

- Capacity of Av. Mario Urteaga, section: Óvalo El Inca to Jr. Dos de Mayo: 1532
- Level of service of Mario Urteaga Avenue, section: El Inca Oval to Dos de Mayo Avenue: E.

Average delay of the traffic signalised intersection: Av. Mario Urteaga - Jr. Urrelo; according to HCM 2010 methodology:

- SE access: 950 s.
- Access NW: 108s.
- SW access:271 s.
- Access NE: 545 s.

Average intersection delay: 653 s.

- Levels of service of all accesses to the traffic-light intersection: Av. Mario Urteaga - Jr. Urrelo; according to HCM 2010 methodology:
- Access SE:F.
- Access NO: F.
- SW access: F: F.

Access NE: F.
Overall intersection level of service: F.
These results corroborate the study hypothesis.

REFERENCES

Ágreda Ríos, L. S., & Parra Alarcón, R. G. (2017). *Comparative analysis of the application of the HCM 2000 and HCM 2010 methodology in a traffic-light section of San Luis Avenue in the metropolitan city of Lima.* Lima: Universidad Peruana de Ciencias Aplicadas.

Alcántara Quispe, M. D. (2018). *Análisis del Nivel de Servicio y Capacidad Vehicular de la Av. San Martín de Porres, ubicada entre ka Av. Atahualpa y la Av. Argentina, aplicando la metodología del HCM 2000.* Cajamarca: Universidad Nacional de Cajamarca.

Arguedas Bernal, C. A., & Mosqueira Grosso, C. E. (2018). *Comprehensive solution proposal for Av. Del Aire between Avenidas Aviación, San Luis and Rosa Toro, through an analysis of vehicular congestion applying the HCM 2010 methodology.* Lima: Universidad Peruana de Ciencias Aplicadas.

Cal, R., Reyes Espíndola, M., & Cárdenas Grisales, J. (1995). *Ingeniería de Tránsito: Fundamentos y aplicaciones.* Mexico: Alfa Omega Gripo Editor.

Chuquihuaccha Mauricio, M. A., & Galván Aldoradín, C. J. (2016). *Propuesta de mejora de congestión vehicular usando metodología del HCM 2010 en las intersecciones de la Av. 26 de noviembre, entre la Av. Pachacútec y la Av. Salvador Allende, en el distrito de V.M.T.* Lima: Universidad Peruana de Ciencias Aplicadas.

Cuenca Poma, P. E. (2014). *Analysis of the capacity and level of service applying HCM methodology on the Loja Zamora road.* Loja: Universidad Católica de Loja.

Fernández De La Maza, J. L. (06 February 2018). *News Week in Spanish.* Retrieved from http://xurl.es/tl8x2

HCM (2010). *Highway Capacity Manual.* Washington DC: Transportation Research Board.

ICG. (2005). *Manual de Diseño Geométrico de Vías Urbanas.* Lima: VCHI.

ICG (2009). *Ingeniería de Tránsito y Seguridad Vial.* Lima: Fondo Editorial ICG.

Llanos Rimarachín, J. M. (2018). *Análisis del Nivel de Servicio de las intersecciones semaforizadas con mayor afluencia de la Av. Hoyos Rubio.* Cajamarca: Universidad Nacional de Cajamarca.

Macera, D. (03 September 2018). Vehicular traffic: Government would save US$500 mlls. if chaos improves. *El Comercio.*

MTC (2006). *Vehicle Classification and Standardisation of Vehicle Registrable Characteristics.* . Lima: El Peruano.

MTC. (2009). *Reglamento Nacional de Tránsito - Código de Tránsito (National Traffic Regulation - Traffic Code).* Lima: Ministerio de Transportes y Comunicaciones.

Ortiz, E. (24 May 2016). *www.lavanguardia.com.* Retrieved from http://xurl.es/0pc4d

Quintero González, J. R. (2017). From the concept of traffic engineering to that of sustainable urban mobility. *Ambiente y desarrollo*, 57-72.

Sabando Santibáñez, I. F. (2017). *Determination of the level of service in urban streets.* Valparaíso: Universidad Técnica Federico Santa María.

ANNEXES

Annex 1

Vehicular flow rate. Direction: Óvalo El Inca - Jr. Dos de Mayo.

I. Datos generales			
Tesis:	ANÁLISIS DE LA CAPACIDAD Y NIVELES DE SERVICIO DE LA AVENIDA MARIO URTEAGA, TRAMO: JR. DOS DE MAYO HASTA EL ÓVALO EL INCA; SEGÚN LA METODOLOGÍA HCM 2010, CAJAMARCA 2018.		
Tesista:	EVER ROJAS HUAMÁN		
Ciudad:	CAJAMARCA		
Calle:	AV. MARIO URTEAGA 191		
Ficha:	TASA DE FLUJO VEHICULAR. TURNO MAÑANA.		
Fecha: 09/11/2018	Hora de inicio: 6:30 hrs	Hora de término: 12:00 hrs	
II. Datos	SENTIDO: ÓVALO EL INCA – JR. DOS DE MAYO		

HORARIO	MOTO TRIMOTO	AUTO	CAMIONETAS PICKUP	CAMIONETAS RURAL Combi	MICRO	BUS B2	TOTAL
6:30 - 6:45	48	14	0	12	0	0	74
6:45 - 7:00	140	37	9	39	5	0	230
7:00 - 7:15	150	20	19	21	9	0	219
7:15 - 7:30	120	42	11	5	3	0	181
7:30 - 7:45	130	59	15	3	0	0	207
7:45 - 8:00	126	50	17	42	1	0	236
8:00 - 8:15	111	40	5	38	3	0	197
8:15 - 8:30	126	54	11	43	4	0	238
8:30 - 8:45	120	17	12	39	6	0	194
8:45 - 9:00	128	36	23	39	4	0	230
9:00 - 9:15	105	32	15	35	5	0	192
9:15 - 9:30	100	39	7	27	5	0	178
9:30 - 9:45	101	69	15	34	2	0	221
9:45 - 10:00	112	41	25	32	2	0	212
10:00 - 10:15	87	51	18	25	4	0	185
10:15 - 10:30	113	55	20	35	3	0	226
10:30 - 10:45	86	43	26	28	3	0	186
10:45 - 11:00	92	63	22	30	5	0	212
11:00 - 11:15	93	38	20	34	3	0	188
11:15 - 11:30	134	57	16	26	5	0	238
11:30 - 11:45	99	42	12	36	4	0	193
11:45 - 12:00	129	11	11	42	3	0	196

I. General data			
Thesis:	ANALYSIS OF THE CAPACITY AND LEVELS OF SERVICE OF AVENIDA MARIO URTEAGA, SECTION: JR. DOS DE MAYO TO EL INCA OVAL; ACCORDING TO THE METHODOLOGY HCM 2010, CAJAMARCA 2018.		
Thesis thesis:	EVER ROJAS HUAMÁN		
City:	CAJAMARCA		
Street:	AV. MARIO URTEAGA 191		
File:	VEHICLE FLOW RATE. MORNING SHIFT.		
Date: 09/11/2018 Start time: 6:30 hrs		End time: 12:00 hrs	
II. Data	DIRECTION: OVALO EL INCA - JR. DOS DE MAYO		

TIME:	MOTORBIKE	AUTO	TRUCKS	MICRO	BUS B2	TOTAL

	TRIMOTO		PICKUP	RURAL Combi			

Annex 2

Vehicle flow rate. Direction: Óvalo El Inca - Jr. Dos de Mayo.

I. General data

Tesis:	**ANÁLISIS DE LA CAPACIDAD Y NIVELES DE SERVICIO DE LA AVENIDA MARIO URTEAGA, TRAMO: JR. DOS DE MAYO HASTA EL ÓVALO EL INCA; SEGÚN LA METODOLOGÍA HCM 2010, CAJAMARCA 2018.**		
Tesista:	**EVER ROJAS HUAMÁN**		
Ciudad:	**CAJAMARCA**		
Calle:	**AV. MARIO URTEAGA 191**		
Ficha:	**TASA DE FLUJO VEHICULAR.** TURNO TARDE.		
Fecha: 09/11/2018		**Hora de inicio: 12:00 hrs**	**Hora de término: 18:00 hrs**

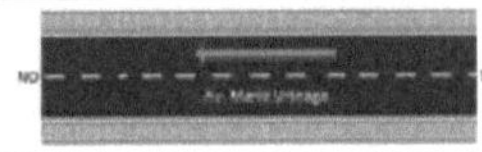

II. Datos	**SENTIDO: ÓVALO EL INCA – JR. DOS DE MAYO**						
HORARIO	**MOTO TRIMOTO**	**AUTO**	**CAMIONETAS**		**MICRO**	**BUS B2**	**TOTAL**
			PICKUP	**RURAL Combi**			
12:00 - 12:15	82	17	14	34	3	0	150
12:15 - 12:30	94	48	23	26	5	0	196
12:30 - 12:45	79	33	28	36	4	0	180
12:45 - 13:00	96	25	26	42	5	0	194
13:00 - 13:15	56	47	23	31	4	0	161
13:15 - 13:30	93	43	24	36	3	0	199
13:30 - 13:45	77	59	19	42	3	0	200
13:45 - 14:00	68	37	10	30	5	0	150
14:00 - 14:15	78	39	13	26	5	0	161
14:15 - 14:30	91	45	10	35	5	0	186
14:30 - 14:45	92	29	13	31	3	0	168
14:45 - 15:00	77	30	21	35	3	0	166
15:00 - 15:15	102	60	19	40	6	0	227
15:15 - 15:30	92	35	16	28	3	0	174
15:30 - 15:45	110	37	28	41	3	0	219
15:45 - 16:00	94	50	20	29	4	0	197
16:00 - 16:15	102	42	31	35	6	0	216
16:15 - 16:30	90	31	23	38	4	0	186
16:30 - 16:45	107	45	25	31	3	0	211
16:45 - 17:00	96	50	20	26	3	0	195

17:00 - 17:15	88	62	23	34	1	0	208
17:15 - 17:30	84	62	22	32	4	0	204
17:30 - 17:45	79	25	26	42	5	0	177
17:45 - 18:00	100	57	30	27	3	0	217

Annex 3

Vehicle flow rate. Direction: Óvalo El Inca - Jr. Dos de Mayo.

I. Datos generales							
Tesis:	**ANÁLISIS DE LA CAPACIDAD Y NIVELES DE SERVICIO DE LA AVENIDA MARIO URTEAGA, TRAMO: JR. DOS DE MAYO HASTA EL ÓVALO EL INCA; SEGÚN LA METODOLOGÍA HCM 2010, CAJAMARCA 2018.**						
Tesista:	**EVER ROJAS HUAMÁN**				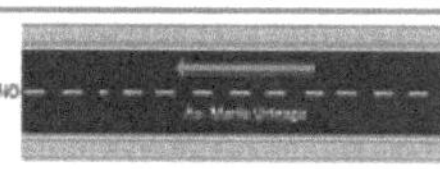		
Ciudad:	**CAJAMARCA**						
Calle:	**AV. MARIO URTEAGA 191**						
Ficha:	**TASA DE FLUJO VEHICULAR.** TURNO NOCHE.						
Fecha: 09/11/2018			**Hora de inicio: 18:00 hrs**		**Hora de término: 21:00 hrs**		
II. Datos	**SENTIDO: ÓVALO EL INCA – JR. DOS DE MAYO**						
HORARIO	**MOTO TRIMOTO**	**AUTO**	**CAMIONETAS**		**MICRO**	**BUS B2**	**TOTAL**
			PICKUP	**RURAL Combi**			
18:00 - 18:15	79	59	21	31	3	0	193
18:15 - 18:30	99	51	31	32	3	0	216
18:30 - 18:45	97	36	19	27	3	0	182
18:45 - 19:00	95	56	36	32	2	0	221
19:00 - 19:15	89	52	19	30	1	0	191
19:15 - 19:30	96	48	30	27	1	0	202
19:30 - 19:45	108	51	33	18	4	0	214
19:45 - 20:00	140	47	27	15	0	0	229
20:00 - 20:15	89	43	17	13	0	0	162
20:15 - 20:30	87	35	15	8	1	0	146
20:30 - 20:45	90	39	11	5	0	0	145
20:45 - 21:00	93	35	20	5	0	0	153

Annex 4 *Vehicle flow rate. Direction: Jr. Dos de Mayo - Óvalo El Inca.*

I. Datos generales			
Tesis:	ANÁLISIS DE LA CAPACIDAD Y NIVELES DE SERVICIO DE LA AVENIDA MARIO URTEAGA, TRAMO: JR. DOS DE MAYO HASTA EL ÓVALO EL INCA; SEGÚN LA METODOLOGÍA HCM 2010, CAJAMARCA 2018.		
Tesista:	EVER ROJAS HUAMÁN		
Ciudad:	CAJAMARCA		
Calle:	AV. MARIO URTEAGA 191		
Ficha:	TASA DE FLUJO VEHICULAR. TURNO MAÑANA.		
Fecha: 09/11/2018		Hora de inicio: 6:30 hrs	Hora de término: 12:00 hrs
II. Datos	SENTIDO: JR. DOS DE MAYO – ÓVALO EL INCA		

HORARIO	MOTO TRIMOTO	AUTO	CAMIONETAS		MICRO	BUS B2	TOTAL
			PICKUP	RURAL Combi			
6:30 - 6:45	29	5	0	1	0	0	35
6:45 - 7:00	96	7	4	5	0	0	112
7:00 - 7:15	132	5	4	6	3	0	150
7:15 - 7:30	105	12	10	3	0	0	130
7:30 - 7:45	145	15	6	5	0	0	171
7:45 - 8:00	118	14	4	10	1	0	147
8:00 - 8:15	89	9	7	12	0	0	117
8:15 - 8:30	86	17	12	16	2	0	133
8:30 - 8:45	95	36	7	8	1	0	147
8:45 - 9:00	104	15	17	6	0	0	142
9:00 - 9:15	71	8	1	6	1	0	87
9:15 - 9:30	76	12	5	7	0	0	100
9:30 - 9:45	71	10	4	3	0	0	88
9:45 - 10:00	80	10	3	7	1	0	101
10:00 - 10:15	72	14	4	3	0	0	93
10:15 - 10:30	63	8	5	9	1	0	86
10:30 - 10:45	82	13	4	5	1	0	105
10:45 - 11:00	73	13	4	7	0	0	97
11:00 - 11:15	87	12	11	5	2	0	117
11:15 - 11:30	116	9	10	11	1	0	147
11:30 - 11:45	107	10	5	24	1	0	147
11:45 - 12:00	97	17	8	13	3	0	138

Annex 5 *Vehicle flow rate. Direction: Jr. Dos de Mayo - Óvalo El Inca.*

I. Datos generales

Tesis:	**ANÁLISIS DE LA CAPACIDAD Y NIVELES DE SERVICIO DE LA AVENIDA MARIO URTEAGA, TRAMO: JR. DOS DE MAYO HASTA EL ÓVALO EL INCA; SEGÚN LA METODOLOGÍA HCM 2010, CAJAMARCA 2018.**
Tesista:	**EVER ROJAS HUAMÁN**
Ciudad:	**CAJAMARCA**
Calle:	**AV. MARIO URTEAGA 191**
Ficha:	**TASA DE FLUJO VEHICULAR.** TURNO TARDE.

Fecha: 09/11/2018	Hora de inicio: 12:00 hrs	Hora de término: 18:00 hrs

II. Datos — **SENTIDO: JR. DOS DE MAYO – ÓVALO EL INCA**

HORARIO	MOTO TRIMOTO	AUTO	CAMIONETAS PICKUP	CAMIONETAS RURAL Combi	MICRO	BUS B2	TOTAL
12:00 - 12:15	93	38	18	23	2	0	174
12:15 - 12:30	108	20	10	14	1	0	153
12:30 - 12:45	92	41	7	24	1	0	165
12:45 - 13:00	83	13	17	10	3	0	126
13:00 - 13:15	74	20	12	11	2	0	119
13:15 - 13:30	74	15	13	10	1	0	113
13:30 - 13:45	77	15	14	7	3	0	116
13:45 - 14:00	69	16	11	8	3	0	107
14:00 - 14:15	59	14	8	12	2	0	95
14:15 - 14:30	61	11	8	12	2	0	94
14:30 - 14:45	76	11	10	8	2	0	107
14:45 - 15:00	66	8	8	7	1	0	90
15:00 - 15:15	68	6	7	10	1	0	92
15:15 - 15:30	84	11	9	8	0	0	112
15:30 - 15:45	78	6	7	9	1	0	101
15:45 - 16:00	83	10	8	8	2	0	111
16:00 - 16:15	87	5	12	10	2	0	116
16:15 - 16:30	82	11	10	8	3	0	114
16:30 - 16:45	74	10	12	9	2	0	107
16:45 - 17:00	82	7	13	7	2	0	111
17:00 - 17:15	91	9	13	8	1	0	122
17:15 - 17:30	67	9	7	11	1	0	95
17:30 - 17:45	83	13	15	10	3	0	124
17:45 - 18:00	76	17	17	14	0	0	124

Annex 6 *Vehicle flow rate. Direction: Jr. Dos de Mayo - El Inca Oval.*

I. Datos generales

Tesis:	**ANÁLISIS DE LA CAPACIDAD Y NIVELES DE SERVICIO DE LA AVENIDA MARIO URTEAGA, TRAMO: JR. DOS DE MAYO HASTA EL ÓVALO EL INCA; SEGÚN LA METODOLOGÍA HCM 2010, CAJAMARCA 2018.**
Tesista:	**EVER ROJAS HUAMÁN**
Ciudad:	**CAJAMARCA**
Calle:	**AV. MARIO URTEAGA 191**
Ficha:	**TASA DE FLUJO VEHICULAR.** TURNO NOCHE.

Fecha: 09/11/2018	**Hora de inicio: 18:00 hrs**	**Hora de término: 21:00 hrs**

II. Datos — **SENTIDO: JR. DOS DE MAYO – ÓVALO EL INCA**

HORARIO	MOTO TRIMOTO	AUTO	CAMIONETAS PICKUP	CAMIONETAS RURAL Combi	MICRO	BUS B2	TOTAL
18:00 - 18:15	56	13	9	3	1	0	82
18:15 - 18:30	101	12	29	4	0	0	146
18:30 - 18:45	99	12	16	14	1	0	142
18:45 - 19:00	72	27	25	9	0	0	133
19:00 - 19:15	70	18	21	11	1	0	121
19:15 - 19:30	48	28	6	4	0	0	86
19:30 - 19:45	112	20	31	2	0	0	165
19:45 - 20:00	129	25	10	14	0	0	178
20:00 - 20:15	77	11	7	12	0	0	107
20:15 - 20:30	63	9	4	5	1	0	82
20:30 - 20:45	59	15	1	6	0	0	81
20:45 - 21:00	56	5	1	4	0	0	66

Annex 7 *Vehicle flow rate. Traffic light intersection. Left turn. Direction: Av. Mario Urteaga - Jr. Urrelo (SE - NO).*

I. Datos generales		
Tesis:	ANÁLISIS DE LA CAPACIDAD Y NIVELES DE SERVICIO DE LA AVENIDA MARIO URTEAGA, TRAMO: JR. DOS DE MAYO HASTA EL ÓVALO EL INCA; SEGÚN LA METODOLOGÍA HCM 2010, CAJAMARCA 2018.	
Tesista:	EVER ROJAS HUAMÁN	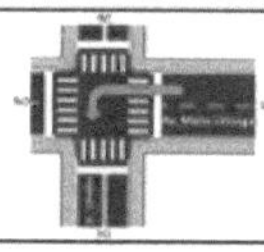
Ciudad:	CAJAMARCA	
Calle:	AV. MARIO URTEAGA 191	
Ficha:	TASA DE FLUJO VEHICULAR. TURNO MAÑANA.	

Fecha: 20/11/2018	Hora de inicio: 6:30 hrs	Hora de término: 12:00 hrs

II. Datos	GIRO A LA IZQUIERDA. SENTIDO: AV. MARIO URTEAGA – JR. URRELO (SE – NO)						
HORARIO	MOTO TRIMOTO	AUTO	CAMIONETAS		MICRO	BUS B2	TOTAL
			PICKUP	RURAL Combi			
6:30 - 6:45	0	5	3	0	0	0	8
6:45 - 7:00	1	10	3	0	0	0	14
7:00 - 7:15	0	12	8	2	0	0	22
7:15 - 7:30	0	10	1	0	0	0	11
7:30 - 7:45	0	12	5	0	0	0	17
7:45 - 8:00	0	10	1	0	0	0	11
8:00 - 8:15	0	20	5	0	0	0	25
8:15 - 8:30	0	15	0	0	0	0	15
8:30 - 8:45	0	14	12	0	0	0	26
8:45 - 9:00	0	18	4	0	0	0	22
9:00 - 9:15	0	16	8	0	0	0	24
9:15 - 9:30	0	14	7	1	0	0	22
9:30 - 9:45	0	13	5	0	0	0	18
9:45 - 10:00	0	12	6	0	0	0	18
10:00 - 10:15	0	17	7	0	0	0	24
10:15 - 10:30	0	14	4	1	0	0	19
10:30 - 10:45	0	16	6	0	0	0	22
10:45 - 11:00	0	16	5	1	0	0	22
11:00 - 11:15	0	44	14	0	0	0	58
11:15 - 11:30	0	40	15	0	0	0	55
11:30 - 11:45	0	35	20	0	0	0	55
11:45 - 12:00	0	39	12	0	0	0	51

Annex 8 Vehicular flow rate. Traffic light intersection. Left turn.
Direction: Av. Mario Urteaga - Jr. Urrelo (SE - NO).

I. Datos generales		
Tesis:	ANÁLISIS DE LA CAPACIDAD Y NIVELES DE SERVICIO DE LA AVENIDA MARIO URTEAGA, TRAMO: JR. DOS DE MAYO HASTA EL ÓVALO EL INCA; SEGÚN LA METODOLOGÍA HCM 2010, CAJAMARCA 2018.	
Tesista:	EVER ROJAS HUAMÁN	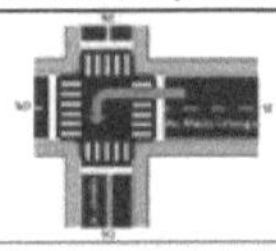
Ciudad:	CAJAMARCA	
Calle:	AV. MARIO URTEAGA 191	
Ficha:	TASA DE FLUJO VEHICULAR. TURNO TARDE.	
Fecha: 20/11/2018	Hora de inicio: 12:00 hrs	Hora de término: 18:00 hrs
II. Datos	GIRO A LA IZQUIERDA. SENTIDO: AV. MARIO URTEAGA – JR. URRELO (SE – NO)	

HORARIO	MOTO TRIMOTO	AUTO	CAMIONETAS PICKUP	CAMIONETAS RURAL Combi	MICRO	BUS B2	TOTAL
12:00 - 12:15	0	48	15	0	0	0	63
12:15 - 12:30	0	46	13	0	0	0	59
12:30 - 12:45	0	38	18	1	0	0	57
12:45 - 13:00	0	40	15	1	0	0	56
13:00 - 13:15	0	16	12	1	0	0	29
13:15 - 13:30	0	17	3	0	0	0	20
13:30 - 13:45	0	25	4	1	0	0	30
13:45 - 14:00	0	14	4	1	0	0	19
14:00 - 14:15	0	8	7	0	0	0	15
14:15 - 14:30	0	17	4	0	0	0	21
14:30 - 14:45	0	10	5	0	0	0	15
14:45 - 15:00	0	15	3	0	0	0	18
15:00 - 15:15	0	17	8	1	0	0	26
15:15 - 15:30	0	12	4	1	0	0	17
15:30 - 15:45	0	12	3	0	0	0	15
15:45 - 16:00	0	11	6	1	0	0	18
16:00 - 16:15	0	12	7	0	0	0	19
16:15 - 16:30	0	13	10	0	0	0	23
16:30 - 16:45	0	19	6	1	0	0	26
16:45 - 17:00	0	10	3	0	0	0	13
17:00 - 17:15	0	11	8	0	0	0	19
17:15 - 17:30	0	24	11	0	0	0	35
17:30 - 17:45	0	26	12	0	0	0	38
17:45 - 18:00	0	22	2	0	0	0	24

Annex 9 Vehicular flow rate. Traffic light intersection. Left turn.

Direction: Av. Mario Urteaga - Jr. Urrelo (SE - NO).

I. Datos generales		
Tesis:	**ANÁLISIS DE LA CAPACIDAD Y NIVELES DE SERVICIO DE LA AVENIDA MARIO URTEAGA, TRAMO: JR. DOS DE MAYO HASTA EL ÓVALO EL INCA; SEGÚN LA METODOLOGÍA HCM 2010, CAJAMARCA 2018.**	
Tesista:	**EVER ROJAS HUAMÁN**	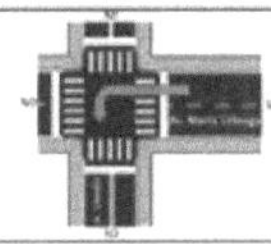
Ciudad:	**CAJAMARCA**	
Calle:	**AV. MARIO URTEAGA 191**	
Ficha:	**TASA DE FLUJO VEHICULAR.** TURNO NOCHE.	
Fecha: 20/11/2018	**Hora de inicio: 18:00 hrs**	**Hora de término: 21:00 hrs**
II. Datos	**GIRO A LA IZQUIERDA. SENTIDO: AV. MARIO URTEAGA – JR. URRELO (SE – NO)**	

HORARIO	MOTO TRIMOTO	AUTO	CAMIONETAS		MICRO	BUS B2	TOTAL
			PICKUP	RURAL Combi			

18:00 - 18:15	0	18	9	0	0	0	27
18:15 - 18:30	0	27	10	0	0	0	37
18:30 - 18:45	0	28	14	0	0	0	42
18:45 - 19:00	0	24	8	1	0	0	33
19:00 - 19:15	0	15	5	1	0	0	21
19:15 - 19:30	0	14	7	1	0	0	22
19:30 - 19:45	0	16	8	0	0	0	24
19:45 - 20:00	0	16	9	0	0	0	25
20:00 - 20:15	0	18	8	1	0	0	27
20:15 - 20:30	0	8	10	0	0	0	18
20:30 - 20:45	0	9	7	1	0	0	17
20:45 - 21:00	0	14	1	1	0	0	16

Annex 10 Vehicular flow rate. Traffic light intersection. Direct passage. Mario Urteaga Av. Direction: SE - NO.

I. Datos generales		
Tesis:	**ANÁLISIS DE LA CAPACIDAD Y NIVELES DE SERVICIO DE LA AVENIDA MARIO URTEAGA, TRAMO: JR. DOS DE MAYO HASTA EL ÓVALO EL INCA; SEGÚN LA METODOLOGÍA HCM 2010, CAJAMARCA 2018.**	
Tesista:	**EVER ROJAS HUAMÁN**	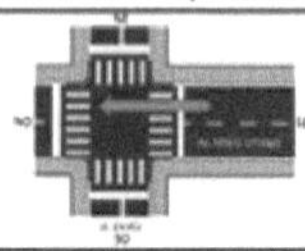
Ciudad:	**CAJAMARCA**	
Calle:	**AV. MARIO URTEAGA 191**	
Ficha:	**TASA DE FLUJO VEHICULAR.** TURNO MAÑANA.	
Fecha: 22/11/2018	**Hora de inicio: 6:30 hrs**	**Hora de término: 12:00 hrs**
II. Datos	**PASO DIRECTO. AV. MARIO URTEAGA. SENTIDO: SE - NO.**	

HORARIO	MOTO TRIMOTO	AUTO	CAMIONETAS PICKUP	CAMIONETAS RURAL Combi	MICRO	BUS B2	TOTAL

6:30 - 6:45	18	8	1	19	2	0	48
6:45 - 7:00	46	8	3	31	6	0	94
7:00 - 7:15	50	9	5	43	3	0	110
7:15 - 7:30	56	9	1	41	6	0	113
7:30 - 7:45	43	8	2	18	5	0	76
7:45 - 8:00	55	10	2	31	2	0	100
8:00 - 8:15	40	13	3	39	1	0	96
8:15 - 8:30	49	17	5	25	2	0	98
8:30 - 8:45	48	9	4	35	6	0	102
8:45 - 9:00	44	6	3	38	3	0	94
9:00 - 9:15	31	17	7	35	6	0	96
9:15 - 9:30	48	14	5	19	2	0	88
9:30 - 9:45	43	14	8	32	5	0	102
9:45 - 10:00	40	18	5	35	6	0	104
10:00 - 10:15	30	14	1	28	3	0	76
10:15 - 10:30	30	16	7	40	6	0	99
10:30 - 10:45	36	21	1	23	1	0	82
10:45 - 11:00	43	13	1	40	6	0	103
11:00 - 11:15	39	9	4	38	1	0	91
11:15 - 11:30	25	7	5	43	6	0	86
11:30 - 11:45	29	15	5	32	3	0	84
11:45 - 12:00	27	11	9	22	5	0	74

Annex 11 Vehicle flow rate. Traffic-light intersection. Direct crossing. Mario Urteaga Av. Direction: SE - NO.

I. Datos generales		
Tesis:	ANÁLISIS DE LA CAPACIDAD Y NIVELES DE SERVICIO DE LA AVENIDA MARIO URTEAGA, TRAMO: JR. DOS DE MAYO HASTA EL ÓVALO EL INCA; SEGÚN LA METODOLOGÍA HCM 2010, CAJAMARCA 2018.	
Tesista:	EVER ROJAS HUAMÁN	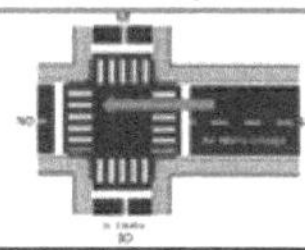
Ciudad:	CAJAMARCA	
Calle:	AV. MARIO URTEAGA 191	
Ficha:	TASA DE FLUJO VEHICULAR. TURNO TARDE.	

Fecha: 22/11/2018	Hora de inicio: 12:00 hrs	Hora de término: 18:00 hrs

II. Datos	PASO DIRECTO. AV. MARIO URTEAGA. SENTIDO: SE - NO.					
HORARIO	MOTO TRIMOTO	AUTO	CAMIONETAS		MICRO	BUS B2	TOTAL

HORARIO	MOTO TRIMOTO	AUTO	CAMIONETAS PICKUP	CAMIONETAS RURAL Combi	MICRO	BUS B2	TOTAL
12:00 - 12:15	29	10	1	28	6	0	74
12:15 - 12:30	47	9	9	29	6	0	100
12:30 - 12:45	25	16	6	31	1	0	79
12:45 - 13:00	41	15	2	39	2	0	99
13:00 - 13:15	43	10	7	34	2	0	96
13:15 - 13:30	35	16	2	35	2	0	90
13:30 - 13:45	41	8	7	44	1	0	101
13:45 - 14:00	48	6	3	27	3	0	87
14:00 - 14:15	56	9	5	25	4	0	99
14:15 - 14:30	58	11	9	28	2	0	108
14:30 - 14:45	61	16	6	33	2	0	118
14:45 - 15:00	69	18	2	27	3	0	119
15:00 - 15:15	38	19	9	19	1	0	86
15:15 - 15:30	48	23	6	25	2	0	104
15:30 - 15:45	47	17	7	34	1	0	106
15:45 - 16:00	49	18	3	37	4	0	111
16:00 - 16:15	49	12	8	23	2	0	94
16:15 - 16:30	48	18	6	20	2	0	94
16:30 - 16:45	46	13	8	39	2	0	108
16:45 - 17:00	37	14	3	22	5	0	81
17:00 - 17:15	29	10	5	31	6	0	81
17:15 - 17:30	44	9	7	23	6	0	89
17:30 - 17:45	44	12	5	25	4	0	90
17:45 - 18:00	46	15	1	26	2	0	90

Annex 12 Vehicle flow rate. Traffic-light intersection. Direct crossing. Mario Urteaga Av. Direction: SE - NO.

I. Datos generales			
Tesis:	ANÁLISIS DE LA CAPACIDAD Y NIVELES DE SERVICIO DE LA AVENIDA MARIO URTEAGA, TRAMO: JR. DOS DE MAYO HASTA EL ÓVALO EL INCA; SEGÚN LA METODOLOGÍA HCM 2010, CAJAMARCA 2018.		
Tesista:	EVER ROJAS HUAMÁN		
Ciudad:	CAJAMARCA		
Calle:	AV. MARIO URTEAGA 191		
Ficha:	TASA DE FLUJO VEHICULAR. TURNO NOCHE.		
Fecha: 22/11/2018	Hora de inicio: 18:00 hrs	Hora de término: 21:00 hrs	
II. Datos	PASO DIRECTO. AV. MARIO URTEAGA. SENTIDO: SE - NO.		

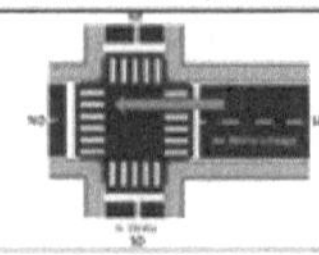

HORARIO	MOTO TRIMOTO	AUTO	CAMIONETAS PICKUP	CAMIONETAS RURAL Combi	MICRO	BUS B2	TOTAL
18:00 - 18:15	47	17	4	32	4	0	104
18:15 - 18:30	28	22	9	35	4	0	98
18:30 - 18:45	31	24	2	39	4	0	100
18:45 - 19:00	42	25	4	38	2	0	111
19:00 - 19:15	27	15	6	38	5	0	91
19:15 - 19:30	37	11	8	17	2	0	75
19:30 - 19:45	33	18	9	21	6	0	87
19:45 - 20:00	35	10	6	25	1	0	77
20:00 - 20:15	12	6	1	11	0	0	30
20:15 - 20:30	28	22	6	8	0	0	64
20:30 - 20:45	38	24	7	4	0	0	73
20:45 - 21:00	37	4	3	5	0	0	49

Annex 13 Vehicle flow rate. Traffic light intersection. Right turn.
Direction: Av. Mario Urteaga - Jr. Urrelo (SE - NW).

I. Datos generales		
Tesis:	**ANÁLISIS DE LA CAPACIDAD Y NIVELES DE SERVICIO DE LA AVENIDA MARIO URTEAGA, TRAMO: JR. DOS DE MAYO HASTA EL ÓVALO EL INCA; SEGÚN LA METODOLOGÍA HCM 2010, CAJAMARCA 2018.**	
Tesista:	**EVER ROJAS HUAMÁN**	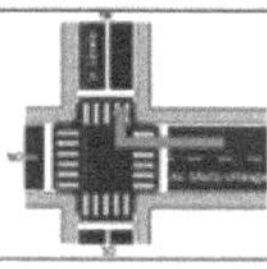
Ciudad:	**CAJAMARCA**	
Calle:	**AV. MARIO URTEAGA 191**	
Ficha:	**TASA DE FLUJO VEHICULAR. TURNO MAÑANA.**	
Fecha: 19/11/2018	**Hora de inicio: 6:30 hrs**	**Hora de término: 12:00 hrs**
II. Datos	**GIRO A LA DERECHA. SENTIDO: AV. MARIO URTEAGA – JR. URRELO (SE – NO)**	

HORARIO	MOTO TRIMOTO	AUTO	CAMIONETAS PICKUP	CAMIONETAS RURAL Combi	MICRO	BUS B2	TOTAL
6:30 - 6:45	32	0	1	2	0	0	35
6:45 - 7:00	78	3	3	2	0	0	86
7:00 - 7:15	90	2	10	0	0	0	102
7:15 - 7:30	90	5	1	1	0	0	97
7:30 - 7:45	80	3	0	0	0	0	83
7:45 - 8:00	57	1	2	0	0	0	60
8:00 - 8:15	64	1	3	0	0	0	68
8:15 - 8:30	66	9	2	0	0	0	77
8:30 - 8:45	60	2	0	0	0	0	62
8:45 - 9:00	72	4	5	0	0	0	81
9:00 - 9:15	66	3	1	0	0	0	70
9:15 - 9:30	52	3	2	1	0	0	58
9:30 - 9:45	49	4	0	1	0	0	54
9:45 - 10:00	59	5	3	1	0	0	68
10:00 - 10:15	48	3	3	0	0	0	54
10:15 - 10:30	72	4	2	0	0	0	78
10:30 - 10:45	63	5	4	0	0	0	72
10:45 - 11:00	54	4	2	0	9	0	69
11:00 - 11:15	40	4	5	0	0	0	49
11:15 - 11:30	58	5	0	1	0	0	64
11:30 - 11:45	72	7	3	1	0	0	83
11:45 - 12:00	63	2	1	1	0	0	67

Annex 14 Vehicular flow rate. Traffic light intersection. Right turn.
Direction: Av. Mario Urteaga - Jr. Urrelo (SE - NO).

I. Datos generales		
Tesis:	ANÁLISIS DE LA CAPACIDAD Y NIVELES DE SERVICIO DE LA AVENIDA MARIO URTEAGA, TRAMO: JR. DOS DE MAYO HASTA EL ÓVALO EL INCA; SEGÚN LA METODOLOGÍA HCM 2010, CAJAMARCA 2018.	
Tesista:	EVER ROJAS HUAMÁN	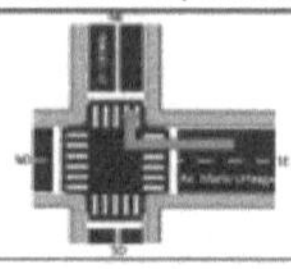
Ciudad:	CAJAMARCA	
Calle:	AV. MARIO URTEAGA 191	
Ficha:	TASA DE FLUJO VEHICULAR. TURNO TARDE.	
Fecha: 19/11/2018	Hora de inicio: 12:00 hrs	Hora de término: 18:00 hrs
II. Datos	GIRO A LA DERECHA. SENTIDO: AV. MARIO URTEAGA – JR. URRELO (SE – NO)	

HORARIO		AUTO	CAMIONETAS		MICRO		TOTAL
	MOTO TRIMOTO		PICKUP	RURAL Combi		BUS B2	
12:00 - 12:15	59	3	1	0	0	0	63
12:15 - 12:30	59	2	2	0	0	0	63
12:30 - 12:45	65	5	1	1	0	0	72
12:45 - 13:00	68	6	3	2	0	0	79
13:00 - 13:15	52	4	6	0	0	0	62
13:15 - 13:30	64	10	4	0	0	0	78
13:30 - 13:45	39	12	5	0	0	0	56
13:45 - 14:00	26	6	2	0	0	0	34
14:00 - 14:15	66	1	1	1	0	0	69
14:15 - 14:30	50	5	2	0	0	0	57
14:30 - 14:45	57	1	3	0	0	0	61
14:45 - 15:00	58	4	0	0	0	0	62
15:00 - 15:15	65	6	1	1	0	0	73
15:15 - 15:30	62	3	2	1	0	0	68
15:30 - 15:45	40	3	2	0	0	0	45
15:45 - 16:00	34	2	2	1	0	0	39
16:00 - 16:15	38	8	1	0	0	0	47
16:15 - 16:30	28	2	2	0	0	0	32
16:30 - 16:45	36	4	2	0	0	0	42
16:45 - 17:00	25	2	2	0	0	0	29
17:00 - 17:15	27	5	1	1	0	0	34
17:15 - 17:30	24	7	2	0	0	0	33
17:30 - 17:45	22	2	2	0	0	0	26
17:45 - 18:00	21	8	4	1	0	0	34

Annex 15 Vehicle flow rate. Traffic light intersection. Right turn. Direction: Av. Mario Urteaga - Jr. Urrelo (SE - NO).

I. Datos generales			
Tesis:	ANÁLISIS DE LA CAPACIDAD Y NIVELES DE SERVICIO DE LA AVENIDA MARIO URTEAGA, TRAMO: JR. DOS DE MAYO HASTA EL ÓVALO EL INCA; SEGÚN LA METODOLOGÍA HCM 2010, CAJAMARCA 2018.		
Tesista:	EVER ROJAS HUAMÁN		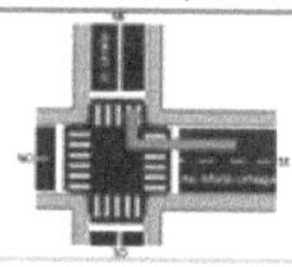
Ciudad:	CAJAMARCA		
Calle:	AV. MARIO URTEAGA 191		
Ficha:	TASA DE FLUJO VEHICULAR. TURNO NOCHE.		
Fecha: 19/11/2018	Hora de inicio: 18:00 hrs	Hora de término: 21:00 hrs	
II. Datos	GIRO A LA DERECHA. SENTIDO: AV. MARIO URTEAGA – JR. URRELO (SE – NO)		

HORARIO	MOTO TRIMOTO	AUTO	CAMIONETAS PICKUP	CAMIONETAS RURAL Combi	MICRO	BUS B2	TOTAL
18:00 - 18:15	31	4	0	0	0	0	35
18:15 - 18:30	34	3	2	1	0	0	40
18:30 - 18:45	17	4	0	0	0	0	21
18:45 - 19:00	27	2	2	0	0	0	31
19:00 - 19:15	22	3	3	1	0	0	29
19:15 - 19:30	26	3	0	0	0	0	29
19:30 - 19:45	22	10	5	0	0	0	37
19:45 - 20:00	27	4	1	1	0	0	33
20:00 - 20:15	41	2	0	0	0	0	43
20:15 - 20:30	37	6	1	1	0	0	45
20:30 - 20:45	32	3	0	0	0	0	35
20:45 - 21:00	29	1	2	0	0	0	32

Annex 16 Vehicle flow rate. Traffic light intersection. Left turn.
Direction: Av. Mario Urteaga - Jr. Urrelo (NW - SE).

Tesis:	**ANÁLISIS DE LA CAPACIDAD Y NIVELES DE SERVICIO DE LA AVENIDA MARIO URTEAGA, TRAMO: JR. DOS DE MAYO HASTA EL ÓVALO EL INCA; SEGÚN LA METODOLOGÍA HCM 2010, CAJAMARCA 2018.**
Tesista:	**EVER ROJAS HUAMÁN**
Ciudad:	**CAJAMARCA**
Calle:	**AV. MARIO URTEAGA 191**
Ficha:	**TASA DE FLUJO VEHICULAR.** TURNO MAÑANA.

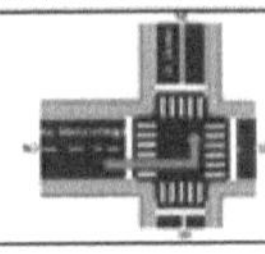

Fecha: 24/11/2018	Hora de inicio: 6:30 hrs	Hora de término: 12:00 hrs

II. Datos — **GIRO A LA IZQUIERDA. SENTIDO: AV. MARIO URTEAGA – JR. URRELO (NO – SE)**

HORARIO	MOTO TRIMOTO	AUTO	CAMIONETAS		MICRO	BUS B2	TOTAL
			PICKUP	RURAL Combi			
6:30 - 6:45	1	1	0	0	0	0	2
6:45 - 7:00	7	1	0	0	0	0	8
7:00 - 7:15	6	4	0	0	0	0	10
7:15 - 7:30	3	2	1	0	0	0	6
7:30 - 7:45	5	0	1	0	0	0	6
7:45 - 8:00	1	0	0	0	0	0	1
8:00 - 8:15	3	1	1	0	0	0	5
8:15 - 8:30	2	3	1	0	0	0	6
8:30 - 8:45	1	3	1	0	0	0	5
8:45 - 9:00	5	3	0	0	0	0	8
9:00 - 9:15	5	2	1	0	0	0	8
9:15 - 9:30	12	1	0	0	0	0	13
9:30 - 9:45	11	1	1	0	0	0	13
9:45 - 10:00	12	1	0	0	0	0	13
10:00 - 10:15	26	0	0	0	0	0	26
10:15 - 10:30	4	0	0	0	0	0	4
10:30 - 10:45	1	1	1	0	0	0	3
10:45 - 11:00	3	1	0	0	0	0	4
11:00 - 11:15	5	1	1	0	0	0	7
11:15 - 11:30	6	0	1	0	0	0	7
11:30 - 11:45	4	3	1	0	0	0	8
11:45 - 12:00	4	2	0	0	0	0	6

Annex 17 Vehicular flow rate. Traffic light intersection. Left turn.
Direction: Av. Mario Urteaga - Jr. Urrelo (NW - SE).

I. Datos generales			
Tesis:	ANÁLISIS DE LA CAPACIDAD Y NIVELES DE SERVICIO DE LA AVENIDA MARIO URTEAGA, TRAMO: JR. DOS DE MAYO HASTA EL ÓVALO EL INCA; SEGÚN LA METODOLOGÍA HCM 2010, CAJAMARCA 2018.		
Tesista:	EVER ROJAS HUAMÁN		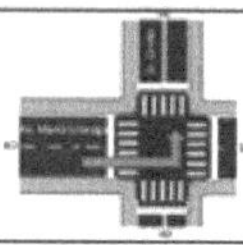
Ciudad:	CAJAMARCA		
Calle:	AV. MARIO URTEAGA 191		
Ficha:	TASA DE FLUJO VEHICULAR. TURNO TARDE.		
Fecha: 24/11/2018	Hora de inicio: 12:00 hrs	Hora de término: 18:00 hrs	
II. Datos	GIRO A LA IZQUIERDA. SENTIDO: AV. MARIO URTEAGA – JR. URRELO (NO – SE)		

HORARIO	MOTO TRIMOTO	AUTO	CAMIONETAS PICKUP	CAMIONETAS RURAL Combi	MICRO	BUS B2	TOTAL
12:00 - 12:15	2	2	0	0	0	0	4
12:15 - 12:30	3	2	0	0	0	0	5
12:30 - 12:45	4	4	0	0	0	0	8
12:45 - 13:00	2	1	1	0	0	0	4
13:00 - 13:15	18	0	3	0	0	0	21
13:15 - 13:30	18	0	1	0	0	0	19
13:30 - 13:45	1	2	0	0	0	0	3
13:45 - 14:00	4	1	0	1	0	0	6
14:00 - 14:15	10	1	0	0	0	0	11
14:15 - 14:30	16	1	0	0	0	0	17
14:30 - 14:45	17	2	0	0	0	0	19
14:45 - 15:00	16	0	0	0	0	0	16
15:00 - 15:15	16	0	0	0	0	0	16
15:15 - 15:30	9	0	0	0	0	0	9
15:30 - 15:45	17	0	0	0	0	0	17
15:45 - 16:00	14	2	1	0	0	0	17
16:00 - 16:15	9	2	0	0	0	0	11
16:15 - 16:30	4	1	0	0	0	0	5
16:30 - 16:45	2	2	0	0	0	0	4
16:45 - 17:00	4	2	0	0	0	0	6
17:00 - 17:15	4	4	1	0	0	0	9
17:15 - 17:30	8	3	1	0	0	0	12
17:30 - 17:45	5	3	1	0	0	0	9
17:45 - 18:00	6	3	1	0	0	0	10

Annex 18 Vehicular flow rate. Traffic light intersection. Left turn.
Direction: Av. Mario Urteaga - Jr. Urrelo (NW - SE).
I. General data Thesis:

I. Datos generales		
Tesis:	**ANÁLISIS DE LA CAPACIDAD Y NIVELES DE SERVICIO DE LA AVENIDA MARIO URTEAGA, TRAMO: JR. DOS DE MAYO HASTA EL ÓVALO EL INCA; SEGÚN LA METODOLOGÍA HCM 2010, CAJAMARCA 2018.**	
Tesista:	**EVER ROJAS HUAMÁN**	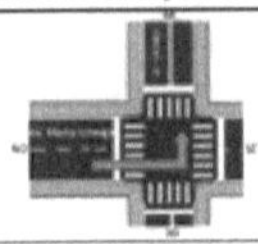
Ciudad:	**CAJAMARCA**	
Calle:	**AV. MARIO URTEAGA 191**	
Ficha:	**TASA DE FLUJO VEHICULAR.** TURNO NOCHE.	
Fecha: 24/11/2018	**Hora de inicio: 18:00 hrs**	**Hora de término: 21:00 hrs**
II. Datos	**GIRO A LA IZQUIERDA. SENTIDO: AV. MARIO URTEAGA – JR. URRELO (NO – SE)**	

HORARIO	MOTO TRIMOTO	AUTO	CAMIONETAS		MICRO	BUS B2	TOTAL
			PICKUP	RURAL Combi			

18:00 - 18:15	2	1	1	0	0	0	4
18:15 - 18:30	3	4	1	0	0	0	8
18:30 - 18:45	4	4	0	0	0	0	8
18:45 - 19:00	4	1	0	0	0	0	5
19:00 - 19:15	2	2	1	0	0	0	5
19:15 - 19:30	1	1	0	0	0	0	2
19:30 - 19:45	2	0	0	0	0	0	2
19:45 - 20:00	1	1	0	0	0	0	2
20:00 - 20:15	2	4	0	0	0	0	6
20:15 - 20:30	3	1	1	0	0	0	5
20:30 - 20:45	1	2	0	0	0	0	3
20:45 - 21:00	0	0	1	0	0	0	1

Annex 19 Vehicular flow rate. Traffic light intersection. Direct crossing. Mario Urteaga Avenue. Direction: NW - SE.

I. Datos generales

Tesis:	**ANÁLISIS DE LA CAPACIDAD Y NIVELES DE SERVICIO DE LA AVENIDA MARIO URTEAGA, TRAMO: JR. DOS DE MAYO HASTA EL ÓVALO EL INCA; SEGÚN LA METODOLOGÍA HCM 2010, CAJAMARCA 2018.**
Tesista:	**EVER ROJAS HUAMÁN**
Ciudad:	**CAJAMARCA**
Calle:	**AV. MARIO URTEAGA 191**
Ficha:	**TASA DE FLUJO VEHICULAR.** TURNO MAÑANA.

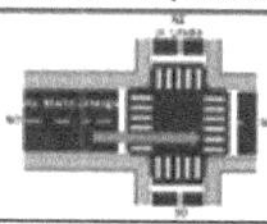

Fecha: 23/11/2018 | **Hora de inicio: 6:30 hrs** | **Hora de término: 12:00 hrs**

II. Datos — **PASO DIRECTO. AV. MARIO URTEAGA. SENTIDO: NO - SE.**

HORARIO	MOTO TRIMOTO	AUTO	CAMIONETAS		MICRO	BUS B2	TOTAL
			PICKUP	RURAL Combi			
6:30 - 6:45	11	2	0	1	0	0	14
6:45 - 7:00	8	3	1	2	0	0	14
7:00 - 7:15	33	2	0	2	0	0	37
7:15 - 7:30	35	7	3	3	0	0	48
7:30 - 7:45	10	9	2	3	0	0	24
7:45 - 8:00	13	11	3	2	0	0	29
8:00 - 8:15	14	9	0	3	0	0	26
8:15 - 8:30	25	8	0	3	0	0	36
8:30 - 8:45	23	10	2	1	0	0	36
8:45 - 9:00	5	7	2	2	0	0	16
9:00 - 9:15	7	11	2	2	0	0	22
9:15 - 9:30	4	9	2	2	0	0	17
9:30 - 9:45	6	11	1	3	0	0	21
9:45 - 10:00	7	11	2	1	0	0	21
10:00 - 10:15	2	8	2	0	0	0	12
10:15 - 10:30	20	11	3	3	0	0	37
10:30 - 10:45	11	11	2	3	0	0	27
10:45 - 11:00	12	4	0	0	0	0	16
11:00 - 11:15	9	12	3	1	0	0	25
11:15 - 11:30	6	11	3	3	0	0	23
11:30 - 11:45	3	7	3	3	0	0	16
11:45 - 12:00	5	3	2	1	0	0	11

Annex 20 Vehicle flow rate. Traffic-light intersection. Direct passage. Mario Urteaga Av. Direction: NO - SE.

I. Datos generales		
Tesis:	ANÁLISIS DE LA CAPACIDAD Y NIVELES DE SERVICIO DE LA AVENIDA MARIO URTEAGA, TRAMO: JR. DOS DE MAYO HASTA EL ÓVALO EL INCA; SEGÚN LA METODOLOGÍA HCM 2010, CAJAMARCA 2018.	
Tesista:	EVER ROJAS HUAMÁN	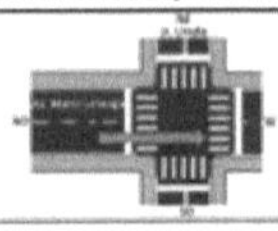
Ciudad:	CAJAMARCA	
Calle:	AV. MARIO URTEAGA 191	
Ficha:	TASA DE FLUJO VEHICULAR. TURNO TARDE.	
Fecha: 23/11/2018	Hora de inicio: 12:00 hrs	Hora de término: 18:00 hrs
II. Datos	PASO DIRECTO. AV. MARIO URTEAGA. SENTIDO: NO - SE.	

HORARIO	MOTO TRIMOTO	AUTO	CAMIONETAS PICKUP	CAMIONETAS RURAL Combi	MICRO	BUS B2	TOTAL
12:00 - 12:15	6	4	3	0	0	0	13
12:15 - 12:30	10	5	0	2	0	0	17
12:30 - 12:45	8	8	1	2	0	0	19
12:45 - 13:00	6	12	3	1	0	0	22
13:00 - 13:15	22	11	1	0	0	0	34
13:15 - 13:30	22	3	2	1	0	0	28
13:30 - 13:45	22	2	1	2	0	0	27
13:45 - 14:00	29	7	1	1	0	0	38
14:00 - 14:15	26	5	1	3	0	0	35
14:15 - 14:30	25	6	2	1	0	0	34
14:30 - 14:45	16	4	1	3	0	0	24
14:45 - 15:00	26	3	0	3	0	0	32
15:00 - 15:15	21	3	1	3	0	0	28
15:15 - 15:30	21	11	3	1	0	0	36
15:30 - 15:45	13	12	3	1	0	0	29
15:45 - 16:00	18	10	0	1	0	0	29
16:00 - 16:15	17	10	1	2	0	0	30
16:15 - 16:30	35	2	3	3	0	0	43
16:30 - 16:45	35	4	3	0	0	0	42
16:45 - 17:00	31	7	2	1	0	0	41
17:00 - 17:15	15	11	0	0	0	0	26
17:15 - 17:30	12	11	3	0	0	0	26
17:30 - 17:45	27	8	1	3	0	0	39
17:45 - 18:00	18	3	2	3	0	0	26

Annex 21 Vehicle flow rate. Traffic-light intersection. Direct passage. Mario Urteaga Av. Direction: NO - SE.

I. Datos generales	
Tesis:	ANÁLISIS DE LA CAPACIDAD Y NIVELES DE SERVICIO DE LA AVENIDA MARIO URTEAGA, TRAMO: JR. DOS DE MAYO HASTA EL ÓVALO EL INCA; SEGÚN LA METODOLOGÍA HCM 2010, CAJAMARCA 2018.
Tesista:	EVER ROJAS HUAMÁN

Ciudad:	CAJAMARCA
Calle:	AV. MARIO URTEAGA 191
Ficha:	TASA DE FLUJO VEHICULAR. TURNO NOCHE.

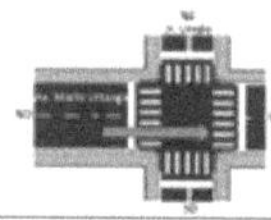

Fecha: 23/11/2018	Hora de inicio: 18:00 hrs	Hora de término: 21:00 hrs

II. Datos	PASO DIRECTO. AV. MARIO URTEAGA. SENTIDO: NO - SE.						
HORARIO	MOTO TRIMOTO	AUTO	CAMIONETAS		MICRO	BUS B2	TOTAL
			PICKUP	RURAL Combi			
18:00 - 18:15	23	10	3	2	0	0	38
18:15 - 18:30	35	7	3	3	0	0	48
18:30 - 18:45	29	11	1	1	0	0	42
18:45 - 19:00	25	4	0	1	0	0	30
19:00 - 19:15	20	10	1	0	0	0	31
19:15 - 19:30	12	3	3	0	0	0	18
19:30 - 19:45	16	3	3	3	0	0	25
19:45 - 20:00	23	2	0	3	0	0	28
20:00 - 20:15	4	8	3	2	0	0	17
20:15 - 20:30	4	8	3	1	0	0	16
20:30 - 20:45	19	5	2	2	0	0	28
20:45 - 21:00	10	7	1	1	0	0	19

Annex 22 Vehicle flow rate. Traffic-light intersection. Right turn. Direction: Av. Mario Urteaga - Jr. Urrelo (NW - SE).

I. Datos generales	
Tesis:	ANÁLISIS DE LA CAPACIDAD Y NIVELES DE SERVICIO DE LA AVENIDA MARIO URTEAGA, TRAMO: JR. DOS DE MAYO HASTA EL ÓVALO EL INCA; SEGÚN LA METODOLOGÍA HCM 2010, CAJAMARCA 2018.
Tesista:	EVER ROJAS HUAMÁN
Ciudad:	CAJAMARCA
Calle:	AV. MARIO URTEAGA 191
Ficha:	TASA DE FLUJO VEHICULAR. TURNO MAÑANA.

Fecha: 21/11/2018	Hora de inicio: 6:30 hrs	Hora de término: 12:00 hrs

II. Datos	GIRO A LA DERECHA. SENTIDO: AV. MARIO URTEAGA – JR. URRELO (NO – SE)						
HORARIO	MOTO TRIMOTO	AUTO	CAMIONETAS		MICRO	BUS B2	TOTAL
			PICKUP	RURAL Combi			
6:30 - 6:45	0	4	1	0	0	0	5
6:45 - 7:00	0	3	0	0	0	0	3
7:00 - 7:15	0	3	1	0	0	0	4
7:15 - 7:30	0	0	3	0	0	0	3
7:30 - 7:45	0	3	1	0	0	0	4
7:45 - 8:00	0	0	1	0	0	0	1
8:00 - 8:15	0	3	1	0	0	0	4
8:15 - 8:30	0	1	0	0	0	0	1
8:30 - 8:45	0	1	3	0	0	0	4
8:45 - 9:00	0	4	0	0	0	0	4
9:00 - 9:15	0	3	0	0	0	0	3
9:15 - 9:30	0	3	3	0	0	0	6
9:30 - 9:45	0	4	0	0	0	0	4
9:45 - 10:00	0	1	2	0	0	0	3
10:00 - 10:15	0	1	3	0	0	0	4
10:15 - 10:30	0	4	2	0	0	0	6
10:30 - 10:45	0	3	3	0	0	0	6
10:45 - 11:00	0	1	1	0	0	0	2
11:00 - 11:15	0	4	2	0	0	0	6
11:15 - 11:30	0	4	1	0	0	0	5
11:30 - 11:45	0	2	0	0	0	0	2
11:45 - 12:00	0	0	1	0	0	0	1

Annex 23 Vehicle flow rate. Traffic light intersection. Right turn.
Direction: Av. Mario Urteaga - Jr. Urrelo (NW - SE).

I. Datos generales		
Tesis:	ANÁLISIS DE LA CAPACIDAD Y NIVELES DE SERVICIO DE LA AVENIDA MARIO URTEAGA, TRAMO: JR. DOS DE MAYO HASTA EL ÓVALO EL INCA; SEGÚN LA METODOLOGÍA HCM 2010, CAJAMARCA 2018.	
Tesista:	EVER ROJAS HUAMÁN	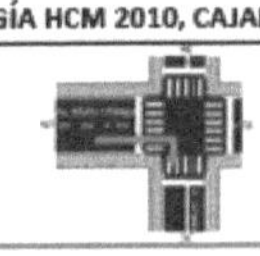
Ciudad:	CAJAMARCA	
Calle:	AV. MARIO URTEAGA 191	
Ficha:	TASA DE FLUJO VEHICULAR. TURNO TARDE.	

Fecha: 21/11/2018			Hora de inicio: 12:00 hrs		Hora de término: 18:00 hrs		
II. Datos	GIRO A LA DERECHA. SENTIDO: AV. MARIO URTEAGA – JR. URRELO (NO – SE)						
HORARIO	MOTO TRIMOTO	AUTO	CAMIONETAS		MICRO	BUS B2	TOTAL
			PICKUP	RURAL Combi			
12:00 - 12:15	0	3	1	0	0	0	4
12:15 - 12:30	0	5	2	0	0	0	7
12:30 - 12:45	0	5	1	0	0	0	6
12:45 - 13:00	0	4	1	0	0	0	5
13:00 - 13:15	5	6	0	0	0	0	11
13:15 - 13:30	4	0	3	0	0	0	7
13:30 - 13:45	7	4	3	0	0	0	14
13:45 - 14:00	4	3	0	0	0	0	7
14:00 - 14:15	0	3	1	0	0	0	4
14:15 - 14:30	0	3	0	0	0	0	3
14:30 - 14:45	0	5	3	0	0	0	8
14:45 - 15:00	0	1	3	0	0	0	4
15:00 - 15:15	0	3	2	0	0	0	5
15:15 - 15:30	0	3	1	0	0	0	4
15:30 - 15:45	0	0	1	0	0	0	1
15:45 - 16:00	0	0	2	0	0	0	2
16:00 - 16:15	0	4	3	0	0	0	7
16:15 - 16:30	0	3	2	0	0	0	5
16:30 - 16:45	0	2	1	0	0	0	3
16:45 - 17:00	0	6	1	0	0	0	7
17:00 - 17:15	0	6	2	0	0	0	8
17:15 - 17:30	0	3	1	0	0	0	4
17:30 - 17:45	0	3	1	0	0	0	4
17:45 - 18:00	0	5	0	0	0	0	5

Annex 24 Vehicle flow rate. Intersection A. Right turn.
Direction: Av. Mario Urteaga - Jr. Urrelo (NW - SE).

I. Datos generales		
Tesis:	**ANÁLISIS DE LA CAPACIDAD Y NIVELES DE SERVICIO DE LA AVENIDA MARIO URTEAGA, TRAMO: JR. DOS DE MAYO HASTA EL ÓVALO EL INCA; SEGÚN LA METODOLOGÍA HCM 2010, CAJAMARCA 2018.**	
Tesista:	**EVER ROJAS HUAMÁN**	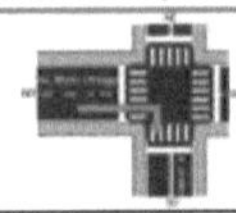
Ciudad:	**CAJAMARCA**	
Calle:	**AV. MARIO URTEAGA 191**	
Ficha:	**TASA DE FLUJO VEHICULAR. TURNO NOCHE.**	
Fecha: 21/11/2018	**Hora de inicio: 18:00 hrs**	**Hora de término: 21:00 hrs**
II. Datos	**GIRO A LA DERECHA. SENTIDO: AV. MARIO URTEAGA – JR. URRELO (NO – SE)**	

HORARIO	MOTO TRIMOTO	AUTO	CAMIONETAS PICKUP	CAMIONETAS RURAL Combi	MICRO	BUS B2	TOTAL
18:00 - 18:15	5	6	0	0	0	0	11
18:15 - 18:30	7	1	2	0	0	0	10
18:30 - 18:45	12	3	0	0	0	0	15
18:45 - 19:00	11	2	0	0	0	0	13
19:00 - 19:15	2	4	0	0	0	0	6
19:15 - 19:30	0	3	0	0	0	0	3
19:30 - 19:45	0	4	2	0	0	0	6
19:45 - 20:00	0	3	1	0	0	0	4
20:00 - 20:15	0	0	0	0	0	0	0
20:15 - 20:30	0	1	1	0	0	0	2
20:30 - 20:45	0	3	0	0	0	0	3
20:45 - 21:00	0	4	0	0	0	0	4

Annex 25

Vehicle flow rate. Traffic light intersection. Left turn.

Direction: Jr. Urrelo - Av. Mario Urteaga (SW - NE).

I. Datos generales		
Tesis:	**ANÁLISIS DE LA CAPACIDAD Y NIVELES DE SERVICIO DE LA AVENIDA MARIO URTEAGA, TRAMO: JR. DOS DE MAYO HASTA EL ÓVALO EL INCA; SEGÚN LA METODOLOGÍA HCM 2010, CAJAMARCA 2018.**	
Tesista:	**EVER ROJAS HUAMÁN**	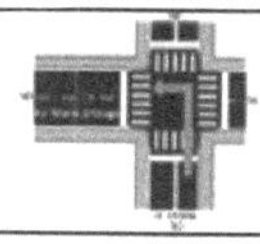
Ciudad:	**CAJAMARCA**	
Calle:	**AV. MARIO URTEAGA 191**	
Ficha:	**TASA DE FLUJO VEHICULAR.** TURNO MAÑANA.	
Fecha: 22/11/2018	**Hora de inicio: 6:30 hrs**	**Hora de término: 12:00 hrs**
II. Datos	**GIRO A LA IZQUIERDA. SENTIDO: JR. URRELO – AV. MARIO URTEGA (SO – NE)**	

HORARIO	MOTO TRIMOTO	AUTO	CAMIONETAS PICKUP	CAMIONETAS RURAL Combi	MICRO	BUS B2	TOTAL
6:30 - 6:45	0	1	0	0	0	0	1
6:45 - 7:00	0	7	0	0	0	0	7
7:00 - 7:15	0	7	0	0	0	0	7
7:15 - 7:30	4	7	1	0	0	0	12
7:30 - 7:45	0	3	3	0	0	0	6
7:45 - 8:00	3	3	0	0	0	0	6
8:00 - 8:15	1	1	2	0	0	0	4
8:15 - 8:30	1	7	3	0	0	0	11
8:30 - 8:45	0	6	2	0	0	0	8
8:45 - 9:00	0	4	2	0	0	0	6
9:00 - 9:15	1	6	1	0	0	0	8
9:15 - 9:30	0	3	2	0	0	0	5
9:30 - 9:45	2	4	2	0	0	0	8
9:45 - 10:00	0	1	0	0	0	0	1
10:00 - 10:15	0	7	0	0	0	0	7
10:15 - 10:30	1	7	3	0	0	0	11
10:30 - 10:45	0	1	2	0	0	0	3
10:45 - 11:00	0	6	3	0	0	0	9
11:00 - 11:15	1	1	2	0	0	0	4
11:15 - 11:30	0	5	0	0	0	0	5
11:30 - 11:45	0	1	0	0	0	0	1
11:45 - 12:00	1	7	0	0	0	0	8

Annex 26 Vehicle flow rate. Intersection with traffic lights. Left turn.
Direction: Jr. Urrelo - Av. Mario Urteaga (SW - NE).

I. Datos generales	
Tesis:	**ANÁLISIS DE LA CAPACIDAD Y NIVELES DE SERVICIO DE LA AVENIDA MARIO URTEAGA, TRAMO: JR. DOS DE MAYO HASTA EL ÓVALO EL INCA; SEGÚN LA METODOLOGÍA HCM 2010, CAJAMARCA 2018.**
Tesista:	**EVER ROJAS HUAMÁN**
Ciudad:	**CAJAMARCA**
Calle:	**AV. MARIO URTEAGA 191**
Ficha:	**TASA DE FLUJO VEHICULAR.** TURNO TARDE.

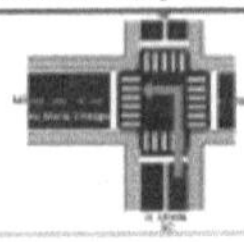

Fecha: 22/11/2018	**Hora de inicio: 12:00 hrs**	**Hora de término: 18:00 hrs**

II. Datos — **GIRO A LA IZQUIERDA. SENTIDO: JR. URRELO – AV. MARIO URTEGA (SO – NE)**

HORARIO	MOTO TRIMOTO	AUTO	CAMIONETAS PICKUP	CAMIONETAS RURAL Combi	MICRO	BUS B2	TOTAL
12:00 - 12:15	0	3	2	0	0	0	5
12:15 - 12:30	0	4	1	0	0	0	5
12:30 - 12:45	1	6	3	0	0	0	10
12:45 - 13:00	0	3	1	0	0	0	4
13:00 - 13:15	2	6	2	3	2	0	15
13:15 - 13:30	4	8	2	4	2	0	20
13:30 - 13:45	2	0	3	0	2	0	7
13:45 - 14:00	3	4	1	0	0	0	8
14:00 - 14:15	6	4	2	0	0	0	12
14:15 - 14:30	1	3	1	1	1	0	7
14:30 - 14:45	0	3	0	0	0	0	3
14:45 - 15:00	3	7	3	0	0	0	13
15:00 - 15:15	0	3	3	0	0	0	6
15:15 - 15:30	0	4	1	0	0	0	5
15:30 - 15:45	1	6	0	0	0	0	7
15:45 - 16:00	0	3	0	0	0	0	3
16:00 - 16:15	0	7	1	0	0	0	8
16:15 - 16:30	1	7	3	0	0	0	11
16:30 - 16:45	0	4	1	0	0	0	5
16:45 - 17:00	0	3	3	0	0	0	6
17:00 - 17:15	2	5	3	0	0	0	10
17:15 - 17:30	0	4	0	0	0	0	4
17:30 - 17:45	1	2	2	0	0	0	5
17:45 - 18:00	0	2	3	0	0	0	5

Annex 27 Vehicle flow rate. Traffic light intersection. Left turn.

Direction: Jr. Urrelo - Av. Mario Urteaga (SW - NE).

I. Datos generales		
Tesis:	ANÁLISIS DE LA CAPACIDAD Y NIVELES DE SERVICIO DE LA AVENIDA MARIO URTEAGA, TRAMO: JR. DOS DE MAYO HASTA EL ÓVALO EL INCA; SEGÚN LA METODOLOGÍA HCM 2010, CAJAMARCA 2018.	
Tesista:	EVER ROJAS HUAMÁN	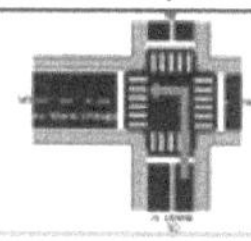
Ciudad:	CAJAMARCA	
Calle:	AV. MARIO URTEAGA 191	
Ficha:	TASA DE FLUJO VEHICULAR. TURNO NOCHE.	
Fecha: 22/11/2018	Hora de inicio: 18:00 hrs	Hora de término: 21:00 hrs
II. Datos	GIRO A LA IZQUIERDA. SENTIDO: JR. URRELO – AV. MARIO URTEGA (SO – NE)	

HORARIO	MOTO TRIMOTO	AUTO	CAMIONETAS PICKUP	CAMIONETAS RURAL Combi	MICRO	BUS B2	TOTAL
18:00 - 18:15	0	3	2	0	0	0	5
18:15 - 18:30	1	1	3	0	0	0	5
18:30 - 18:45	0	7	1	0	0	0	8
18:45 - 19:00	0	2	2	0	0	0	4
19:00 - 19:15	0	4	2	0	0	0	6
19:15 - 19:30	1	7	2	0	0	0	10
19:30 - 19:45	0	6	2	0	0	0	8
19:45 - 20:00	0	7	1	0	0	0	8
20:00 - 20:15	2	4	1	0	0	0	7
20:15 - 20:30	0	2	2	0	0	0	4
20:30 - 20:45	0	3	0	0	0	0	3
20:45 - 21:00	1	2	2	0	0	0	5

Annex 28 Vehicular flow rate. Traffic light intersection. Direct passage. Jr. Urrelo. Direction: SW - NE.

I. Datos generales			
Tesis:	**ANÁLISIS DE LA CAPACIDAD Y NIVELES DE SERVICIO DE LA AVENIDA MARIO URTEAGA, TRAMO: JR. DOS DE MAYO HASTA EL ÓVALO EL INCA; SEGÚN LA METODOLOGÍA HCM 2010, CAJAMARCA 2018.**		
Tesista:	**EVER ROJAS HUAMÁN**	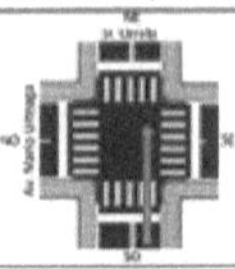	
Ciudad:	**CAJAMARCA**		
Calle:	**AV. MARIO URTEAGA 191**		
Ficha:	**TASA DE FLUJO VEHICULAR.** TURNO MAÑANA.		
Fecha: 21/11/2018	**Hora de inicio: 6:30 hrs**	**Hora de término: 12:00 hrs**	
II. Datos	**PASO DIRECTO. JR. URRELO. SENTIDO: SO - NE.**		

HORARIO	MOTO TRIMOTO	AUTO	CAMIONETAS PICKUP	CAMIONETAS RURAL Combi	MICRO	BUS B2	TOTAL
6:30 - 6:45	5	9	1	2	0	0	17
6:45 - 7:00	17	13	2	0	0	0	32
7:00 - 7:15	12	17	10	2	0	0	41
7:15 - 7:30	11	19	18	3	0	0	51
7:30 - 7:45	17	13	14	3	0	0	47
7:45 - 8:00	9	19	10	2	0	0	40
8:00 - 8:15	7	16	7	1	0	0	31
8:15 - 8:30	4	19	6	2	0	0	31
8:30 - 8:45	2	11	3	0	0	0	16
8:45 - 9:00	4	19	16	3	0	0	42
9:00 - 9:15	0	16	11	1	0	0	28
9:15 - 9:30	6	10	7	3	0	0	26
9:30 - 9:45	7	11	15	3	0	0	36
9:45 - 10:00	4	18	9	2	0	0	33
10:00 - 10:15	6	17	4	2	0	0	29
10:15 - 10:30	3	19	11	1	0	0	34
10:30 - 10:45	6	18	12	0	0	0	36
10:45 - 11:00	2	10	6	0	0	0	18
11:00 - 11:15	2	19	12	1	0	0	34
11:15 - 11:30	5	11	15	2	0	0	33
11:30 - 11:45	7	13	5	0	0	0	25
11:45 - 12:00	2	17	10	2	0	0	31

Annex 29 Vehicle flow rate. Traffic light intersection. Direct crossing. Jr. Urrelo. Direction: SW - NE.

I. General data Thesis:

I. Datos generales		
Tesis:	ANÁLISIS DE LA CAPACIDAD Y NIVELES DE SERVICIO DE LA AVENIDA MARIO URTEAGA, TRAMO: JR. DOS DE MAYO HASTA EL ÓVALO EL INCA; SEGÚN LA METODOLOGÍA HCM 2010, CAJAMARCA 2018.	
Tesista:	EVER ROJAS HUAMÁN	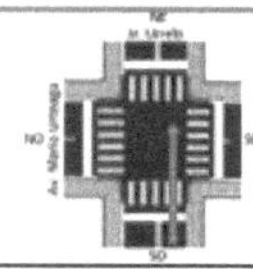
Ciudad:	CAJAMARCA	
Calle:	AV. MARIO URTEAGA 191	
Ficha:	TASA DE FLUJO VEHICULAR. TURNO TARDE.	
Fecha: 21/11/2018	Hora de inicio: 12:00 hrs	Hora de término: 18:00 hrs
II. Datos	PASO DIRECTO. JR. URRELO. SENTIDO: SO - NE.	

HORARIO	MOTO TRIMOTO	AUTO	CAMIONETAS		MICRO	BUS B2	TOTAL
			PICKUP	RURAL Combi			

12:00 - 12:15	5	16	6	2	0	0	29
12:15 - 12:30	3	24	5	1	0	0	33
12:30 - 12:45	9	17	4	1	0	0	31
12:45 - 13:00	3	20	7	2	0	0	32
13:00 - 13:15	21	36	12	7	0	0	76
13:15 - 13:30	11	22	18	1	1	0	53
13:30 - 13:45	16	21	4	5	0	0	46
13:45 - 14:00	15	26	8	2	0	0	51
14:00 - 14:15	1	13	16	2	0	0	32
14:15 - 14:30	4	24	11	3	0	0	42
14:30 - 14:45	5	25	12	0	0	0	42
14:45 - 15:00	0	24	5	1	0	0	30
15:00 - 15:15	9	26	16	3	0	0	54
15:15 - 15:30	6	24	9	0	0	0	39
15:30 - 15:45	6	12	19	2	0	0	39
15:45 - 16:00	3	23	8	2	0	0	36
16:00 - 16:15	6	20	15	0	0	0	41
16:15 - 16:30	3	22	8	3	0	0	36
16:30 - 16:45	6	17	7	2	0	0	32
16:45 - 17:00	3	18	7	3	0	0	31
17:00 - 17:15	4	13	3	1	0	0	21
17:15 - 17:30	2	19	16	3	0	0	40
17:30 - 17:45	4	15	18	0	0	0	37
17:45 - 18:00	3	15	12	0	0	0	30

Annex 30 Vehicle flow rate. Traffic signalised intersection. Direct crossing. Jr. Urrelo. Direction: SW - NE.

I. Datos generales	
Tesis:	ANÁLISIS DE LA CAPACIDAD Y NIVELES DE SERVICIO DE LA AVENIDA MARIO URTEAGA, TRAMO: JR. DOS DE MAYO HASTA EL ÓVALO EL INCA; SEGÚN LA METODOLOGÍA HCM 2010, CAJAMARCA 2018.
Tesista:	EVER ROJAS HUAMÁN
Ciudad:	CAJAMARCA
Calle:	AV. MARIO URTEAGA 191
Ficha:	TASA DE FLUJO VEHICULAR. TURNO NOCHE.

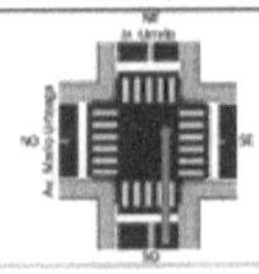

Fecha: 21/11/2018	Hora de inicio: 18:00 hrs	Hora de término: 21:00 hrs

II. Datos	PASO DIRECTO. JR. URRELO. SENTIDO: SO - NE.

HORARIO	MOTO TRIMOTO	AUTO	CAMIONETAS		MICRO	BUS B2	TOTAL
			PICKUP	RURAL Combi			
18:00 - 18:15	18	18	7	2	0	0	45
18:15 - 18:30	7	23	8	2	1	0	41
18:30 - 18:45	9	38	12	3	0	0	62
18:45 - 19:00	10	26	11	3	0	0	50
19:00 - 19:15	1	27	16	2	0	0	46
19:15 - 19:30	5	24	9	1	0	0	39
19:30 - 19:45	1	28	9	1	0	0	39
19:45 - 20:00	0	11	4	1	0	0	16
20:00 - 20:15	4	21	2	2	0	0	29
20:15 - 20:30	1	30	3	0	0	0	34
20:30 - 20:45	6	17	4	0	0	0	27
20:45 - 21:00	1	26	4	0	0	0	31

Annex 31 Vehicle flow rate. Traffic light intersection. Right turn. Direction: Jr. Urrelo - Av. Mario Urteaga (SW - NE).

I. Datos generales	
Tesis:	ANÁLISIS DE LA CAPACIDAD Y NIVELES DE SERVICIO DE LA AVENIDA MARIO URTEAGA, TRAMO: JR. DOS DE MAYO HASTA EL ÓVALO EL INCA; SEGÚN LA METODOLOGÍA HCM 2010, CAJAMARCA 2018.
Tesista:	EVER ROJAS HUAMÁN
Ciudad:	CAJAMARCA
Calle:	AV. MARIO URTEAGA 191
Ficha:	TASA DE FLUJO VEHICULAR. TURNO MAÑANA.

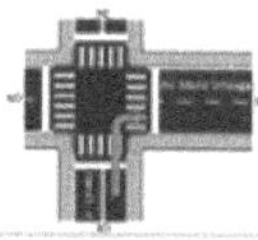

Fecha: 21/11/2018	Hora de inicio: 6:30 hrs	Hora de término: 12:00 hrs

II. Datos — GIRO A LA DERECHA. SENTIDO: JR. URRELO – AV. MARIO URTEAGA (SO – NE)

HORARIO	MOTO TRIMOTO	AUTO	CAMIONETAS		MICRO	BUS B2	TOTAL
			PICKUP	RURAL Combi			
6:30 - 6:45	4	2	2	0	0	0	8
6:45 - 7:00	4	2	0	0	0	0	6
7:00 - 7:15	4	1	1	0	0	0	6
7:15 - 7:30	2	0	0	0	0	0	2
7:30 - 7:45	3	2	2	0	0	0	7
7:45 - 8:00	0	2	0	0	0	0	2
8:00 - 8:15	3	0	1	0	0	0	4
8:15 - 8:30	3	1	0	0	0	0	4
8:30 - 8:45	0	1	2	0	0	0	3
8:45 - 9:00	4	2	1	0	0	0	7
9:00 - 9:15	3	2	1	0	0	0	6
9:15 - 9:30	0	1	2	0	0	0	3
9:30 - 9:45	2	1	1	0	0	0	4
9:45 - 10:00	4	0	0	0	0	0	4
10:00 - 10:15	4	2	2	0	0	0	8
10:15 - 10:30	1	1	0	0	0	0	2
10:30 - 10:45	1	0	2	0	0	0	3
10:45 - 11:00	4	2	1	0	0	0	7
11:00 - 11:15	1	0	1	0	0	0	2
11:15 - 11:30	2	2	0	0	0	0	4
11:30 - 11:45	1	1	2	0	0	0	4
11:45 - 12:00	4	2	0	0	0	0	6

Annex 32

Vehicle flow rate. Traffic light intersection. Right turn.

Direction: Jr. Urrelo - Av. Mario Urteaga (SW - NE).

I. Datos generales		
Tesis:	**ANÁLISIS DE LA CAPACIDAD Y NIVELES DE SERVICIO DE LA AVENIDA MARIO URTEAGA, TRAMO: JR. DOS DE MAYO HASTA EL ÓVALO EL INCA; SEGÚN LA METODOLOGÍA HCM 2010, CAJAMARCA 2018.**	
Tesista:	**EVER ROJAS HUAMÁN**	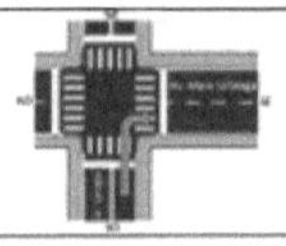
Ciudad:	**CAJAMARCA**	
Calle:	**AV. MARIO URTEAGA 191**	
Ficha:	**TASA DE FLUJO VEHICULAR.** TURNO TARDE.	
Fecha: 21/11/2018	**Hora de inicio: 12:00 hrs**	**Hora de término: 18:00 hrs**
II. Datos	**GIRO A LA DERECHA. SENTIDO: JR. URRELO – AV. MARIO URTEAGA (SO – NE)**	

HORARIO	MOTO TRIMOTO	AUTO	CAMIONETAS PICKUP	CAMIONETAS RURAL Combi	MICRO	BUS B2	TOTAL
12:00 - 12:15	1	1	0	0	0	0	2
12:15 - 12:30	4	1	1	0	0	0	6
12:30 - 12:45	3	2	0	0	0	0	5
12:45 - 13:00	4	2	2	0	0	0	8
13:00 - 13:15	7	5	5	0	0	0	17
13:15 - 13:30	2	1	1	0	0	0	4
13:30 - 13:45	5	2	1	0	0	0	8
13:45 - 14:00	6	1	0	0	0	0	7
14:00 - 14:15	0	0	2	0	0	0	2
14:15 - 14:30	2	1	1	0	0	0	4
14:30 - 14:45	2	2	2	0	0	0	6
14:45 - 15:00	2	0	2	0	0	0	4
15:00 - 15:15	3	2	1	0	0	0	6
15:15 - 15:30	4	0	0	0	0	0	4
15:30 - 15:45	3	2	1	0	0	0	6
15:45 - 16:00	4	2	2	0	0	0	8
16:00 - 16:15	3	0	0	0	0	0	3

16:15 - 16:30	4	2	0	0	0	0	6
16:30 - 16:45	4	1	0	0	0	0	5
16:45 - 17:00	2	1	1	0	0	0	4
17:00 - 17:15	4	2	2	0	0	0	8
17:15 - 17:30	1	0	2	0	0	0	3
17:30 - 17:45	3	1	2	0	0	0	6
17:45 - 18:00	3	2	2	0	0	0	7

Annex 33 Vehicle flow rate. Intersection with traffic lights. Right turn. Direction: Jr. Urrelo - Av. Mario Urteaga (SW - NE).

I. Datos generales		
Tesis:	ANÁLISIS DE LA CAPACIDAD Y NIVELES DE SERVICIO DE LA AVENIDA MARIO URTEAGA, TRAMO: JR. DOS DE MAYO HASTA EL ÓVALO EL INCA; SEGÚN LA METODOLOGÍA HCM 2010, CAJAMARCA 2018.	
Tesista:	EVER ROJAS HUAMÁN	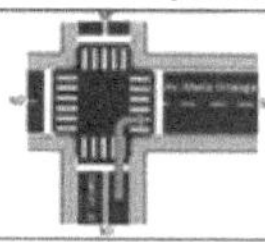
Ciudad:	CAJAMARCA	
Calle:	AV. MARIO URTEAGA 191	
Ficha:	TASA DE FLUJO VEHICULAR. TURNO NOCHE.	
Fecha: 21/11/2018	Hora de inicio: 18:00 hrs	Hora de término: 21:00 hrs

II. Datos	GIRO A LA DERECHA. SENTIDO: JR. URRELO – AV. MARIO URTEAGA (SO – NE)						
HORARIO	MOTO TRIMOTO	AUTO	CAMIONETAS		MICRO	BUS	TOTAL
			PICKUP	RURAL Combi		B2	
18:00 - 18:15	1	1	0	0	0	0	2
18:15 - 18:30	0	1	1	0	0	0	2
18:30 - 18:45	2	1	0	0	0	0	3
18:45 - 19:00	2	4	2	0	0	0	8
19:00 - 19:15	0	1	0	0	0	0	1
19:15 - 19:30	0	2	1	0	0	0	3
19:30 - 19:45	2	1	1	0	0	0	4
19:45 - 20:00	2	0	1	0	0	0	3
20:00 - 20:15	0	1	0	0	0	0	1
20:15 - 20:30	4	1	0	0	0	0	5
20:30 - 20:45	0	1	0	0	0	0	1
20:45 - 21:00	1	2	0	0	0	0	3

Annex 34 Vehicle flow rate. Intersection with traffic lights. Left turn.
Direction: Jr. Urrelo - Av. Mario Urteaga (NE - SW).

I. Datos generales		
Tesis:	ANÁLISIS DE LA CAPACIDAD Y NIVELES DE SERVICIO DE LA AVENIDA MARIO URTEAGA, TRAMO: JR. DOS DE MAYO HASTA EL ÓVALO EL INCA; SEGÚN LA METODOLOGÍA HCM 2010, CAJAMARCA 2018.	
Tesista:	EVER ROJAS HUAMÁN	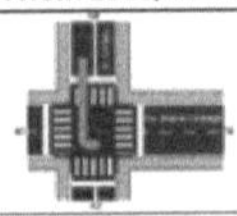
Ciudad:	CAJAMARCA	
Calle:	AV. MARIO URTEAGA 191	
Ficha:	TASA DE FLUJO VEHICULAR. TURNO MAÑANA.	
Fecha: 20/11/2018	Hora de inicio: 6:30 hrs	Hora de término: 12:00 hrs
II. Datos	GIRO A LA IZQUIERDA. SENTIDO: JR. URRELO – AV. MARIO URTEAGA (NE – SO)	

HORARIO	MOTO TRIMOTO	AUTO	CAMIONETAS		MICRO	BUS B2	TOTAL
			PICKUP	RURAL Combi			
6:30 - 6:45	13	1	1	0	0	0	15
6:45 - 7:00	26	3	1	0	0	0	30
7:00 - 7:15	35	0	1	0	0	0	36
7:15 - 7:30	35	9	2	0	0	0	46
7:30 - 7:45	13	2	0	0	0	0	15
7:45 - 8:00	28	6	1	0	0	0	35
8:00 - 8:15	0	0	0	0	0	0	0
8:15 - 8:30	0	0	0	0	0	0	0
8:30 - 8:45	0	0	1	0	0	0	1
8:45 - 9:00	2	1	1	0	0	0	4
9:00 - 9:15	0	0	1	0	0	0	1
9:15 - 9:30	0	4	1	0	0	0	5
9:30 - 9:45	2	0	2	0	0	0	4
9:45 - 10:00	2	1	3	0	0	0	6
10:00 - 10:15	1	5	1	0	0	0	7
10:15 - 10:30	0	3	3	1	0	0	7
10:30 - 10:45	0	5	0	0	0	0	5
10:45 - 11:00	2	2	1	0	0	0	5
11:00 - 11:15	16	2	0	0	0	0	18
11:15 - 11:30	20	0	1	0	0	0	21
11:30 - 11:45	23	2	0	0	0	0	25
11:45 - 12:00	26	2	2	0	0	0	30

Annex 35 Vehicle flow rate. Intersection with traffic lights. Left turn. Direction: Jr. Urrelo - Av. Mario Urteaga (NE - SW).

I. Datos generales	
Tesis:	**ANÁLISIS DE LA CAPACIDAD Y NIVELES DE SERVICIO DE LA AVENIDA MARIO URTEAGA, TRAMO: JR. DOS DE MAYO HASTA EL ÓVALO EL INCA; SEGÚN LA METODOLOGÍA HCM 2010, CAJAMARCA 2018.**
Tesista:	**EVER ROJAS HUAMÁN**
Ciudad:	**CAJAMARCA**
Calle:	**AV. MARIO URTEAGA 191**
Ficha:	**TASA DE FLUJO VEHICULAR.** TURNO TARDE.

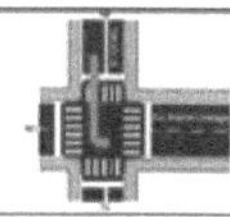

Fecha: 20/11/2018	**Hora de inicio: 12:00 hrs**	**Hora de término: 18:00 hrs**

II. Datos	**GIRO A LA IZQUIERDA. SENTIDO: JR. URRELO – AV. MARIO URTEAGA (NE – SO)**

HORARIO	MOTO TRIMOTO	AUTO	CAMIONETAS		MICRO	BUS B2	TOTAL
			PICKUP	RURAL Combi			
12:00 - 12:15	17	2	0	0	0	0	19
12:15 - 12:30	14	3	2	0	0	0	19
12:30 - 12:45	24	3	1	0	0	0	28
12:45 - 13:00	13	1	1	0	0	0	15
13:00 - 13:15	12	1	1	0	0	0	14
13:15 - 13:30	12	4	0	0	0	0	16
13:30 - 13:45	8	2	3	0	0	0	13
13:45 - 14:00	7	2	3	0	0	0	12
14:00 - 14:15	1	2	2	0	0	0	5
14:15 - 14:30	0	1	2	0	0	0	3
14:30 - 14:45	1	4	2	0	0	0	7
14:45 - 15:00	1	5	0	0	0	0	6
15:00 - 15:15	2	5	2	0	0	0	9
15:15 - 15:30	2	3	1	0	0	0	6
15:30 - 15:45	1	3	3	0	0	0	7
15:45 - 16:00	1	3	0	1	0	0	5
16:00 - 16:15	1	0	1	0	0	0	2
16:15 - 16:30	1	2	0	0	0	0	3
16:30 - 16:45	2	5	1	0	0	0	8
16:45 - 17:00	0	2	3	0	0	0	5
17:00 - 17:15	2	3	1	0	0	0	6
17:15 - 17:30	2	3	0	0	0	0	5
17:30 - 17:45	25	3	0	0	0	0	28
17:45 - 18:00	21	4	0	0	0	0	25

Annex 36 Vehicle flow rate. Intersection with traffic lights. Left turn.

Direction: Jr. Urrelo - Av. Mario Urteaga (NE - SW).

I. Datos generales			
Tesis:	**ANÁLISIS DE LA CAPACIDAD Y NIVELES DE SERVICIO DE LA AVENIDA MARIO URTEAGA, TRAMO: JR. DOS DE MAYO HASTA EL ÓVALO EL INCA; SEGÚN LA METODOLOGÍA HCM 2010, CAJAMARCA 2018.**		
Tesista:	**EVER ROJAS HUAMÁN**	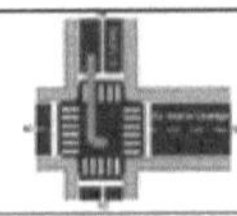	
Ciudad:	**CAJAMARCA**		
Calle:	**AV. MARIO URTEAGA 191**		
Ficha:	**TASA DE FLUJO VEHICULAR.** TURNO NOCHE.		
Fecha: 20/11/2018	**Hora de inicio: 18:00 hrs**	**Hora de término: 21:00 hrs**	
II. Datos	**GIRO A LA IZQUIERDA. SENTIDO: JR. URRELO – AV. MARIO URTEAGA (NE – SO)**		

HORARIO	MOTO TRIMOTO	AUTO	CAMIONETAS PICKUP	CAMIONETAS RURAL Combi	MICRO	BUS B2	TOTAL
18:00 - 18:15	22	2	1	0	0	0	25
18:15 - 18:30	27	4	5	0	0	0	36
18:30 - 18:45	17	4	2	0	0	0	23
18:45 - 19:00	30	1	0	0	0	0	31
19:00 - 19:15	15	2	3	0	0	0	20
19:15 - 19:30	11	1	0	0	0	0	12
19:30 - 19:45	17	1	1	0	0	0	19
19:45 - 20:00	25	4	1	0	0	0	30
20:00 - 20:15	10	4	2	0	0	0	16
20:15 - 20:30	4	5	2	0	0	0	11
20:30 - 20:45	2	1	2	0	0	0	5
20:45 - 21:00	1	4	1	0	0	0	6

Annex 37 Vehicle flow rate. Traffic light intersection. Direct passage. Jr. Urrelo. Direction: NE - SW.

I. Datos generales		
Tesis:	ANÁLISIS DE LA CAPACIDAD Y NIVELES DE SERVICIO DE LA AVENIDA MARIO URTEAGA, TRAMO: JR. DOS DE MAYO HASTA EL ÓVALO EL INCA; SEGÚN LA METODOLOGÍA HCM 2010, CAJAMARCA 2018.	
Tesista:	EVER ROJAS HUAMÁN	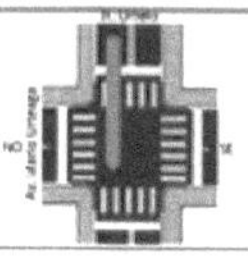
Ciudad:	CAJAMARCA	
Calle:	AV. MARIO URTEAGA 191	
Ficha:	TASA DE FLUJO VEHICULAR. TURNO MAÑANA.	
Fecha: 20/11/2018	Hora de inicio: 6:30 hrs	Hora de término: 12:00 hrs
II. Datos	PASO DIRECTO. JR. URRELO. SENTIDO: NE - SO.	

HORARIO	MOTO TRIMOTO	AUTO	CAMIONETAS PICKUP	CAMIONETAS RURAL Combi	MICRO	BUS B2	TOTAL
6:30 - 6:45	0	5	1	0	0	0	6
6:45 - 7:00	3	8	2	0	0	0	13
7:00 - 7:15	3	12	7	1	0	0	23
7:15 - 7:30	2	3	5	0	0	0	10
7:30 - 7:45	0	4	2	0	0	0	6
7:45 - 8:00	0	8	6	2	0	0	16
8:00 - 8:15	0	6	1	0	1	0	8
8:15 - 8:30	0	18	2	0	0	0	20
8:30 - 8:45	0	19	6	0	0	0	25
8:45 - 9:00	0	19	5	0	0	0	24
9:00 - 9:15	0	15	0	0	0	0	15
9:15 - 9:30	0	17	5	0	0	0	22
9:30 - 9:45	1	16	0	1	0	0	18
9:45 - 10:00	0	14	3	1	0	0	18
10:00 - 10:15	0	8	1	0	0	0	9
10:15 - 10:30	1	4	6	0	0	0	11
10:30 - 10:45	1	9	7	0	0	0	17
10:45 - 11:00	1	11	7	2	0	0	21
11:00 - 11:15	5	11	1	0	0	0	17
11:15 - 11:30	0	4	3	0	0	0	7
11:30 - 11:45	0	8	8	0	0	0	16
11:45 - 12:00	0	12	1	0	0	0	13

Annex 38 Vehicle flow rate. Traffic signalised intersection. Direct crossing. Jr. Urrelo. Direction: NE - SW.

I. Datos generales	
Tesis:	ANÁLISIS DE LA CAPACIDAD Y NIVELES DE SERVICIO DE LA AVENIDA MARIO URTEAGA, TRAMO: JR. DOS DE MAYO HASTA EL ÓVALO EL INCA; SEGÚN LA METODOLOGÍA HCM 2010, CAJAMARCA 2018.
Tesista:	EVER ROJAS HUAMÁN
Ciudad:	CAJAMARCA
Calle:	AV. MARIO URTEAGA 191

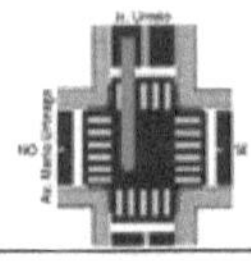

Ficha:	TASA DE FLUJO VEHICULAR. TURNO TARDE.
Fecha: 20/11/2018	Hora de inicio: 12:00 hrs — Hora de término: 18:00 hrs
II. Datos	PASO DIRECTO. JR. URRELO. SENTIDO: NE - SO.

HORARIO	MOTO TRIMOTO	AUTO	CAMIONETAS PICKUP	CAMIONETAS RURAL Combi	MICRO	BUS B2	TOTAL
12:00 - 12:15	0	6	5	0	0	0	11
12:15 - 12:30	0	11	1	0	0	0	12
12:30 - 12:45	0	11	4	0	0	0	15
12:45 - 13:00	0	13	1	0	0	0	14
13:00 - 13:15	0	8	3	0	0	0	11
13:15 - 13:30	0	9	5	0	0	0	14
13:30 - 13:45	1	8	2	1	1	0	13
13:45 - 14:00	0	9	6	0	0	0	15
14:00 - 14:15	0	15	4	0	0	0	19
14:15 - 14:30	1	16	3	0	0	0	20
14:30 - 14:45	0	17	3	0	0	0	20
14:45 - 15:00	1	16	3	0	0	0	20
15:00 - 15:15	1	18	6	1	0	0	26
15:15 - 15:30	1	13	7	1	0	0	22
15:30 - 15:45	1	17	5	0	0	0	23
15:45 - 16:00	1	7	5	0	0	0	13
16:00 - 16:15	0	9	4	0	0	0	13
16:15 - 16:30	1	7	2	0	0	0	10
16:30 - 16:45	0	11	6	0	0	0	17
16:45 - 17:00	1	14	5	0	0	0	20

17:00 - 17:15	1	11	1	0	0	0	13
17:15 - 17:30	1	11	6	0	0	0	18
17:30 - 17:45	0	20	8	0	0	0	28
17:45 - 18:00	0	23	5	0	0	0	28

Annex 39 Vehicle flow rate. Traffic-light intersection. Direct crossing. Jr. Urrelo. Direction: NE - SW.

I. Datos generales

Tesis:	**ANÁLISIS DE LA CAPACIDAD Y NIVELES DE SERVICIO DE LA AVENIDA MARIO URTEAGA, TRAMO: JR. DOS DE MAYO HASTA EL ÓVALO EL INCA; SEGÚN LA METODOLOGÍA HCM 2010, CAJAMARCA 2018.**
Tesista:	**EVER ROJAS HUAMÁN**
Ciudad:	**CAJAMARCA**
Calle:	**AV. MARIO URTEAGA 191**
Ficha:	**TASA DE FLUJO VEHICULAR.** TURNO NOCHE.

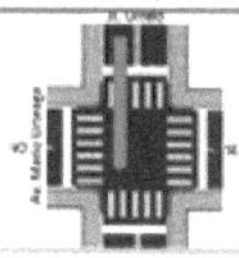

Fecha: 20/11/2018 | **Hora de inicio: 18:00 hrs** | **Hora de término: 21:00 hrs**

II. Datos — **PASO DIRECTO. JR. URRELO. SENTIDO: NE - SO.**

HORARIO	MOTO TRIMOTO	AUTO	CAMIONETAS PICKUP	CAMIONETAS RURAL Combi	MICRO	BUS B2	TOTAL
18:00 - 18:15	0	9	9	1	0	0	19
18:15 - 18:30	1	14	6	1	0	0	22
18:30 - 18:45	0	11	6	0	0	0	17
18:45 - 19:00	0	23	5	1	0	0	29
19:00 - 19:15	0	9	7	0	0	0	16
19:15 - 19:30	0	16	2	0	1	0	19
19:30 - 19:45	0	6	1	0	0	0	7
19:45 - 20:00	2	16	4	0	0	0	22
20:00 - 20:15	0	13	3	0	0	0	16
20:15 - 20:30	0	16	2	0	0	0	18
20:30 - 20:45	0	6	6	0	0	0	12
20:45 - 21:00	0	5	6	0	0	0	11

Annex 40 Vehicle flow rate. Traffic light intersection. Right turn.
Direction: Jr. Urrelo - Av. Mario Urteaga (NE - SW).

I. Datos generales							
Tesis:	ANÁLISIS DE LA CAPACIDAD Y NIVELES DE SERVICIO DE LA AVENIDA MARIO URTEAGA, TRAMO: JR. DOS DE MAYO HASTA EL ÓVALO EL INCA; SEGÚN LA METODOLOGÍA HCM 2010, CAJAMARCA 2018.						
Tesista:	EVER ROJAS HUAMÁN				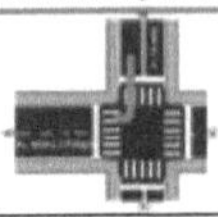		
Ciudad:	CAJAMARCA						
Calle:	AV. MARIO URTEAGA 191						
Ficha:	TASA DE FLUJO VEHICULAR. TURNO MAÑANA.						
Fecha: 22/11/2018			Hora de inicio: 6:30 hrs		Hora de término: 12:00 hrs		
II. Datos	GIRO A LA DERECHA. SENTIDO: JR. URRELO – AV. MARIO URTEAGA (NE – SO)						
HORARIO	MOTO TRIMOTO	AUTO	CAMIONETAS		MICRO	BUS B2	TOTAL
			PICKUP	RURAL Combi			

6:30 - 6:45	0	3	1	0	0	0	4
6:45 - 7:00	18	3	0	0	0	0	21
7:00 - 7:15	22	2	1	0	0	0	25
7:15 - 7:30	40	1	1	1	0	0	43
7:30 - 7:45	16	2	1	0	0	0	19
7:45 - 8:00	32	0	0	0	0	0	32
8:00 - 8:15	22	0	1	0	0	0	23
8:15 - 8:30	32	2	1	0	0	0	35
8:30 - 8:45	19	3	0	1	0	0	23
8:45 - 9:00	38	0	1	0	0	0	39
9:00 - 9:15	34	2	1	0	0	0	37
9:15 - 9:30	22	0	0	0	0	0	22
9:30 - 9:45	31	3	1	0	0	0	35
9:45 - 10:00	35	2	1	0	0	0	38
10:00 - 10:15	12	0	0	0	0	0	12
10:15 - 10:30	14	1	1	0	0	0	16
10:30 - 10:45	10	0	0	0	0	0	10
10:45 - 11:00	24	3	0	0	0	0	27
11:00 - 11:15	31	3	0	0	0	0	34
11:15 - 11:30	28	0	1	0	0	0	29
11:30 - 11:45	13	3	0	0	0	0	16
11:45 - 12:00	36	1	0	0	0	0	37

Annex 41

Vehicular flow rate. Intersection with traffic lights. Right turn.

Direction: Jr. Urrelo - Av. Mario Urteaga (NE - SW).

I. Datos generales		
Tesis:	**ANÁLISIS DE LA CAPACIDAD Y NIVELES DE SERVICIO DE LA AVENIDA MARIO URTEAGA, TRAMO: JR. DOS DE MAYO HASTA EL ÓVALO EL INCA; SEGÚN LA METODOLOGÍA HCM 2010, CAJAMARCA 2018.**	
Tesista:	**EVER ROJAS HUAMÁN**	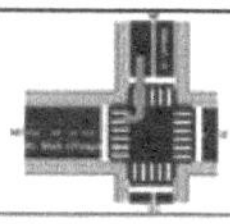
Ciudad:	**CAJAMARCA**	
Calle:	**AV. MARIO URTEAGA 191**	
Ficha:	**TASA DE FLUJO VEHICULAR.** TURNO TARDE.	
Fecha: 22/11/2018	**Hora de inicio: 12:00 hrs**	**Hora de término: 18:00 hrs**
II. Datos	**GIRO A LA DERECHA. SENTIDO: JR. URRELO – AV. MARIO URTEAGA (NE – SO)**	

HORARIO	MOTO TRIMOTO	AUTO	CAMIONETAS		MICRO	BUS	TOTAL
			PICKUP	RURAL Combi		B2	
						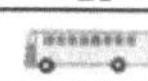	
12:00 - 12:15	25	0	1	0	0	0	26
12:15 - 12:30	15	2	0	0	0	0	17
12:30 - 12:45	10	2	1	0	0	0	13
12:45 - 13:00	24	2	0	0	0	0	26
13:00 - 13:15	34	10	4	0	0	0	48
13:15 - 13:30	28	3	3	0	0	0	34
13:30 - 13:45	26	5	3	0	0	0	34
13:45 - 14:00	19	2	0	0	0	0	21
14:00 - 14:15	31	2	2	0	0	0	35
14:15 - 14:30	49	4	8	3	0	0	64
14:30 - 14:45	49	5	7	0	0	0	61
14:45 - 15:00	45	4	2	0	0	0	51
15:00 - 15:15	29	4	0	0	0	0	33
15:15 - 15:30	23	3	0	0	0	0	26
15:30 - 15:45	33	4	0	0	0	0	37
15:45 - 16:00	24	4	0	0	0	0	28
16:00 - 16:15	28	1	1	0	0	0	30
16:15 - 16:30	32	3	0	0	0	0	35
16:30 - 16:45	30	1	0	0	0	0	31
16:45 - 17:00	34	3	0	0	0	0	37
17:00 - 17:15	34	4	0	0	0	0	38
17:15 - 17:30	8	2	1	0	0	0	11
17:30 - 17:45	29	0	0	0	0	0	29
17:45 - 18:00	9	1	0	0	0	0	10

Annex 42 Vehicle flow rate. Intersection with traffic lights. Right turn.
Direction: Jr. Urrelo - Av. Mario Urteaga (NE - SW).

I. Datos generales		
Tesis:	ANÁLISIS DE LA CAPACIDAD Y NIVELES DE SERVICIO DE LA AVENIDA MARIO URTEAGA, TRAMO: JR. DOS DE MAYO HASTA EL ÓVALO EL INCA; SEGÚN LA METODOLOGÍA HCM 2010, CAJAMARCA 2018.	
Tesista:	EVER ROJAS HUAMÁN	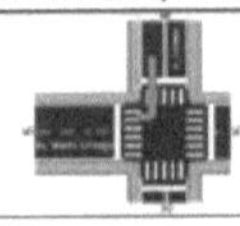
Ciudad:	CAJAMARCA	
Calle:	AV. MARIO URTEAGA 191	
Ficha:	TASA DE FLUJO VEHICULAR. TURNO NOCHE.	
Fecha: 22/11/2018	Hora de inicio: 18:00 hrs	Hora de término: 21:00 hrs
II. Datos	GIRO A LA DERECHA. SENTIDO: JR. URRELO – AV. MARIO URTEAGA (NE – SO)	

HORARIO	MOTO TRIMOTO	AUTO	CAMIONETAS		MICRO	BUS B2	TOTAL
			PICKUP	RURAL Combi			

18:00 - 18:15	9	2	1	0	0	0	12
18:15 - 18:30	7	3	0	0	0	0	10
18:30 - 18:45	14	4	1	0	0	0	19
18:45 - 19:00	13	2	1	0	0	0	16
19:00 - 19:15	20	4	1	0	0	0	25
19:15 - 19:30	17	1	0	0	0	0	18
19:30 - 19:45	7	3	1	0	0	0	11
19:45 - 20:00	9	2	1	0	0	0	12
20:00 - 20:15	8	1	1	0	0	0	10
20:15 - 20:30	7	1	0	0	0	0	8
20:30 - 20:45	0	0	0	0	0	0	0
20:45 - 21:00	1	0	0	0	0	0	1

FSC
www.fsc.org
MIX
Papier | Fördert gute Waldnutzung
FSC® C083411

Kolibri 360 GmbH
Ferdinand-Jühlke-Straße 7
99095 Erfurt, Deutschland
produktsicherheit@kolibri360.de

Druck:
CPI Druckdienstleistungen GmbH
im Auftrag der
Kolibri 360 GmbH
Ferdinand-Jühlke-Str. 7
99095 Erfurt